普通高等教育“十二五”规划双语系列教材

NUMERICAL SIMULATION OF GROUNDWATER FLOW—FINITE-DIFFERENCE METHOD

地下水流数值模拟——有限差分法

郑秀清　陈军锋　刘萍　编著

内 容 提 要

本教材是国内地下水流数值模拟的一本双语教材，内容包括基本理论、Visual MODFLOW 软件使用和实践应用三部分，知识点系统全面。第一部分是地下水流数值模拟的基础理论，讲述了有关基本概念、基本定律、基本微分方程、数学模型的建立和有限差分方法等。第二部分讲述了 Visual MODFLOW 软件的使用步骤，包括水流模型的输入、运行和输出。第三部分为基于 Visual MODFLOW 软件的地下水流数值模拟应用实例，包括地下水库截渗墙建设的数值模拟和水源地保护区划分的数值模拟。

编者根据多年的教学经验和水文与水资源工程专业本科生双语课程"Numerical Simulation of Groundwater Flow"的教学示范总结，并融合了基本的水文地质学基础知识和地下水动力学理论，既注重强化学生的基础理论，又侧重培养学生的动手能力，教材特色鲜明，可作为高等院校水文与水资源工程、地下水科学与工程、市政工程和环境工程等相关专业本科生和研究生的教材。

图书在版编目（CIP）数据

地下水流数值模拟 : 有限差分法 = Numerical simulation of groundwater flow:finite-difference method : 汉英对照 / 郑秀清, 陈军锋, 刘萍编著. -- 北京 : 中国水利水电出版社, 2014.2
普通高等教育"十二五"规划双语系列教材
ISBN 978-7-5170-1784-4

Ⅰ. ①地… Ⅱ. ①郑… ②陈… ③刘… Ⅲ. ①地下水－数值模拟－有限差分法－高等学校－双语教学－教材－汉、英 Ⅳ. ①P641.2

中国版本图书馆CIP数据核字(2014)第046656号

书　名	普通高等教育"十二五"规划双语系列教材 **NUMERICAL SIMULATION OF GROUNDWATER FLOW —FINITE－DIFFERENCE METHOD** **地下水流数值模拟——有限差分法**
作　者	郑秀清　陈军锋　刘萍　编著
出版发行	中国水利水电出版社 （北京市海淀区玉渊潭南路1号D座　100038） 网址：www.waterpub.com.cn E－mail：sales@waterpub.com.cn 电话：（010）68367658（发行部）
经　售	北京科水图书销售中心（零售） 电话：（010）88383994、63202643、68545874 全国各地新华书店和相关出版物销售网点
排　版	中国水利水电出版社微机排版中心
印　刷	北京瑞斯通印务发展有限公司
规　格	184mm×260mm　16开本　8.25印张　280千字
版　次	2014年2月第1版　2014年2月第1次印刷
印　数	0001—2000册
定　价	**29.00**元

ABSTRACT

This bilingual textbook about numerical simulation of groundwater flow, can be divided into three parts including basic theory, Visual MODFLOW software introduction and practical application. The first part describes the basic concepts, fundamental law, basic differential equations, establishment of mathematical model and the finite-difference method. The second part presents the tutorial of Visual MODFLOW software, including the input, run and output of groundwater flow model. Based on Visual MODFLOW software, the numerical simulations of groundwater flow application case are recommended in the third part, including numerical simulation of impervious wall construction and protection zones delineation.

On the basis of many years of teaching experience and summary of bilingual teaching on "NUMERICAL SIMULATION OF GROUNDWATER FLOW" which is setting for Hydrology and Water Resources Engineering specialty, we have integrated basic hydrogeology knowledge and groundwater dynamics theory in the book. This distinctive textbook focuses on not only strengthening basic theory of hydrogeology but also cultivating students' practical ability. So this bilingual textbook can be used by undergraduates and graduates who are major in Hydrology and Water Resources Engineering, Groundwater Science and Engineering, Municipal Engineering and Environmental Engineering.

前 言

地下水是宝贵的自然资源，也是人类赖以生存的物质基础。随着社会经济的快速发展，地下水不合理开发利用和污染问题日益严重，已成为水利、环境、地矿、城建等领域重要的研究对象。地下水流数值模拟是研究地下水流运动规律的强有力工具。

近年来，双语教学作为高校提升教学水平和学生综合素质的教学手段，得到相关教育部门的高度重视，正在逐步走进课堂。此外，国内使用的地下水流的数值模拟软件大多是英文界面和英文说明书，从模拟软件的教学和实践应用角度来看，也迫切需要正式出版一本“地下水流数值模拟”方面的双语教材。编者多年来的双语教学实践表明，课堂教学采用中英文双语教学，不仅能够提高学生的专业知识和地下水流数值模拟的能力，而且可以提高学生的专业英语阅读水平和写作水平。因此，为了适应部分高校教学改革发展，以及改善教材不能与时俱进的局面，适时编写出版了本教材。

全书共分 6 章，第 1 章、第 4 章和第 5 章由太原理工大学陈军锋执笔，第 3 章由太原理工大学郑秀清执笔，第 2 章和第 6 章由太原理工大学刘萍执笔，全书由郑秀清教授统稿，部分图件由臧红飞博士负责完成。在此，我们还要特别感谢研究生任霞、冯晓曦和雷俊琴等对本教材的顺利出版付出的辛勤劳动。

我们对所有为本书修改、出版付出辛勤劳动的同志致以衷心的感谢。本书不当之处在所难免，恳请读者给予指正。

作 者

2014 年 2 月

PREFACE

Groundwater is a precious natural resources and a basic material for human survival. As the social economy advances rapidly, unreasonable exploitation and serious pollution problems of groundwater has attracted more attention from researchers who are engaged in Water Conservation, Environment, Mining, and Civil Construction and other domains. Groundwater flow numerical simulation is viewed as a powerful tool to inquiry the law of groundwater movement.

In recent years, bilingual teaching, which can improve teaching level and students' comprehensive quality, has been paid more attention by Education Departments and is entering into classrooms step by step as a teaching means. In addition, the software of groundwater flow numerical simulation in our country mostly has English interface and instructions. It is urgent to compile and to publish a bilingual textbook about groundwater flow numerical simulation for the sake of teaching and the practical application of simulation software.

Many years experiences of bilingual teaching have tell us that using both English and Chinese in classroom teaching can not only promote the students' professional knowledge and the ability of groundwater flow numerical simulation, but also enhance the students' capacity for English reading and writing. Therefore, in order to adapt to the college teaching reform and development, and improve the situation that the teaching material can't keep pace with the times, we publish this book.

The book is divided into six chapters. The first, fourth and fifth chapters are written by Chen Junfeng of Taiyuan University of Technology; the third chapter is written by Professor Zheng Xiuqing of Taiyuan University of Technology; the second and sixth chapters are written by Liu

Ping of Taiyuan University of Technology, and Professor Zheng Xiuqing does the final compilation and editing. Some of the figures are finished by Dr Zang Hongfei. Many thanks to the graduate students Ren Xia, Feng Xiaoxi and Lei Junqin for doing a lot of hard work for the publication of this book!

We express our heartfelt thanks to all comrades who have been engaged in this book. Improprieties in this book are inevitable, and we will appreciate for your correction.

Authors

February, 2014

Contents

Chapter 1 Introduction

§ 1.1 Brief of Groundwater

Groundwater is the most precious natural resources for a country. Throughout history, people around the world have used groundwater as a source of drinking water and even today, more than half of the world's population depend on groundwater for survival. Many countries have relied on groundwater for generations, with little thought of using it up or contaminating it. All countries hope to improve their economy by increasing industrial or agricultural production. This goal results in an increase in water use and the potential for contamination. Water managers once believed groundwater was a pure resource, isolated from sources of contamination. However, groundwater contamination has emerged as a major environmental problem in many countries. The public's attention has been drawn to the problem because of the many incidents of groundwater contamination. In the United States, this situation has led to expensive groundwater cleanup, the establishment of groundwater protection laws and environmental protection programs.

1.1.1 Importance of Groundwater

Water is vital to man's existence. Early human civilization was centered around springs and streams. The waterhole was the forerunner of a well. Early man copied animals and found that water could be obtained by digging cavities in wadi beds or damp places.

Hidden beneath much of the land surface are groundwater reservoirs-open spaces or voids in rocks which store the largest volume of liquid freshwater on Earth. More than 20 percent of Earth's freshwater resources are here. It is the source of drinking water for most of the world's rural population and is a vital resources, especially in arid areas and on islands, where it may be the only source for potable water. It is essential to maintain soil moisture for crops, lake levels, streamflow, and wetlands. Groundwater from wells and springs is the major source of bottled mineral drinking water.

Groundwater is present in permeable rocks beneath most land areas. Where it occurs, it should be considered a possible alternative supply to surface water. As a water supply, groundwater has the following potential advantages.

(1) Compared with the high costs of constructing surface water reservoirs, the drillings of wells is relatively inexpensive and can be phased in over a period of specified time to meet increasing demand.

(2) The environmental impact of a well is minor.

(3) Many aquifers have large storage capacities so that increasing demand for water during extended droughts can be met more easily.

(4) Groundwater is usually of good chemical and bacterial quality and is unlikely to require treatment other than precautionary chlorination.

Contrary to commonly-held belief, groundwater is a renewable resource. In many parts of the world, groundwater supplies are continually replaced by rainfall and ensuing infiltration although in arid and semiarid regions, the recover rate may be slow or periodic and the recover amount is small. It is important to realize that all artificial withdrawal of groundwater is at the expense of natural discharge. By careful and knowledgeable management, however, the effects of pumping can be controlled to minimize adverse environmental consequences. In the long term, water levels will only decline if groundwater withdrawals exceed the local rates of replenishment. This custom, called groundwater "mining", is more likely to occur in semiarid regions with limited or no replenishment.

1.1.2 Groundwater Resources Management

Groundwater is a renewable resource, Therefore, in order to achieve long-term benefits, not only the sufficient assessment of groundwater potential but also the high efficiency of water use is required. There are many countries in the world where groundwater is one of the major sources of drinking water. With the increasing development of the groundwater resources and the growing impacts of human activities on the aquifers, problems such as declines of groundwater heads and deterioration of groundwater quality have been observed in many places in last decades. Sustainable development strategy and integrated groundwater resources management must be developed and implemented to guarantee the right of use of the limited water resources for our future generations.

To formulate technically reasonable groundwater resources management polices, decision makers always ask questions like:

(1) How long can an aquifer maintain the current rate of groundwater abstraction? What is the safety yield that the aquifer can sustain the continuous abstraction?

(2) What is the capture zone of a water supply well field? What is the most likely pathway of contaminants from domestic wastewater and leaches from solid waste disposal sites? What are the chances that the pollutants from those sources would arrive at water supply wells? And how long it takes? In order to protect the well fields from pollution, a protection zone should be delineated. What is the size of the protection zone?

Providing answers to these questions involves the understanding of the behavior of groundwater flow system and the prediction of the system's response to any stresses. Numerical simulation, a useful tool for groundwater management and protection, is always used in solving these problems. Understanding of groundwater system characteristics is the precondition of building mathematical model. The groundwater flow problems and the groundwater quality problems can be solved efficiently by numerical simulation. In order to quantitatively calculate the available exploitation quantity of groundwater and provide basis for rational development and utilization of water resource, the numerical simulation method is used to establish a three-dimensional numerical simulation model. An optimal scheme is determined to ensure the normal operation of the water source.

§ 1. 2　Process of Groundwater Modeling

The process of groundwater modeling involves a number of different steps and the essential steps are shown as Figure 1. 1.

1. *Defining purpose*

Groundwater models are usually applied for predicting the consequences of the proposed actions such as groundwater development scenarios or waste disposal. Models can be used for analyzing groundwater flow system by assembling and organizing field data and formulating ideas about dynamics of flow systems. Models can be also used for studying processes in generic geologic settings like river-aquifer systems. It is essential to identify clearly the purpose of modeling so that the needs of modeling efforts and accuracy are determined. The purpose of modeling also decides on the dimensionality and time dependency of a model.

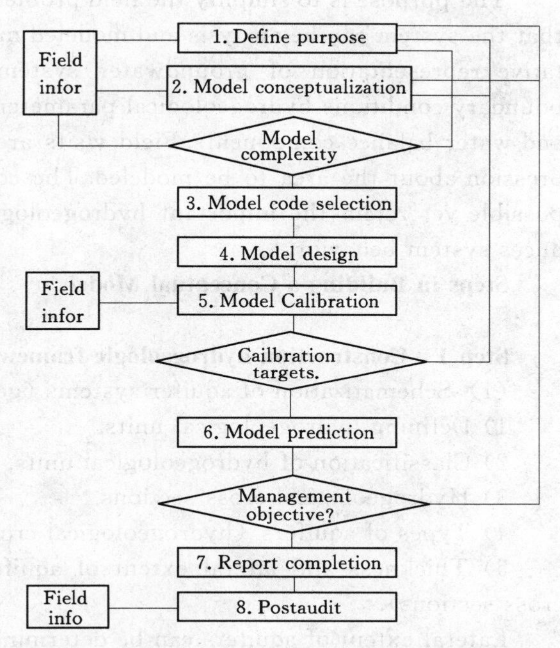

Figure 1. 1　Process of groundwater modeling

Answers to the following questions will help in the determination of the types of model applications and the levels of modeling efforts:

(1) Will the model be used for the prediction of system's response, analysis of flow systems, or study of the processes in a certain generic geologic settings? (Really necessary to build a model?)

(2) What questions do you want the model to answer? (Questions to be answered by the model.)

(3) Can an analytical model provides the answer or must be a numerical model is constructed? (Analytical or numerical model?)

Examples of prediction the consequences of a proposed action:

(1) Groundwater development scenarios: Extension and magnitude of the cone of depression around a pumping station.

(2) Groundwater pollution: Plume of groundwater contaminants from a waste disposal site.

(3) Interaction between groundwater and environment impacts of a reservoir on groundwater level.

Example of interpretation:

(1) Framework for assemble field data synthesizing field data; testing assumptions about the system; indicating the further field work.

(2) Flow system analysis: pathlines, flow rate, pattern of recharge and discharge boundary conditions.

(3) Sensitivity analysis: Identification of important system parameters.

2. *Building conceptual model*

The purpose is to simplify the field problem and organize the associated field data so that the system can be analysis and modeled more readily. Conceptual model is a quantitative representation of groundwater systems in terms of aquifer-aquitard layers, boundary conditions hydrogeological parameters, hydrological stresses, flow patterns, and water balance components. Field visits are necessary to gain the modeler first impression about the area to be modeled. The conceptual model is simplified as much as possible yet retain the important hydrogeologic condition so that it adequately reproduces system behavior.

Steps in Building a Conceptual Model

Step 1 Construction hydrogeologic framework.

(1) Schematization of aquifer systems (geological cross sections).

1) Defining hydrogeological units.

2) Classification of hydrogeological units.

3) Hydrogeological cross sections.

4) Types of aquifers (hydrogeological cross sections).

5) Thickness and lateral extent of aquifers and confining beds (hydrogeological cross sections).

Lateral extent of aquifers can be determined from cross sections and then projected on the map; Natural hydrogeological boundaries are boundaries of the extent of aquifers; Construction of contour maps of groundwater level, elevation of bottoms of aquifers and confining beds; Aquifer thickness can be calculated from contour maps, or directly calculated from cross sections; Construction of isopach map of aquifers and confining beds.

(2) Boundaries of aquifer systems (hydrogeological cross sections).

1) Types of boundaries.

Physical boundaries (fixed)—Impermeable rocks; Impermeable faults; Large bodies of surface water.

Hydraulic boundaries (movable)—Groundwater divides; Streamlines.

2) Mathematical representation.

3) Setting boundaries.

Hydrogeological boundaries—Impermeable rocks; Impermeable faults; Large rivers, lakes, and oceans; Regional groundwater divides.

Distant boundaries—Artificial boundaries for transient simulation where head and flow are not influenced by the stresses.

Hydraulic boundaries—Groundwater divides; Streamlines; Groundwater head contour line.

4) Simulating boundaries.

Specified head boundaries—River; Lake; Ocean; Water level.

Specified flow boundaries—Seepage to stream, spring flow, underflow, seepage to/from bedrocks, local hydraulic boundaries.

No-flow boundaries—Impermeable bedrock, impermeable fault zone, seepage a groundwater divide, a streamline, a freshwater/saltwater interface.

Head-dependent flow boundaries—Leakage to/from river, lake, reservoir.

(3) Hydrogeologic parameters.

1) Parameters—Hydraulic conductivity, K; Transmissivity, $T=Km$; Storage coefficient, S_s; Specific yield, S_y; Porosity, n; and so on.

2) Pumping tests.

3) Laboratory tests.

4) Empirical data.

(4) Extent and rate of areal recharge (precipitation, irrigation).

(5) Extent and rate of areal discharge (evapotranspiration).

(6) Locations and rate of wells (discharge/recharge).

(7) Spatial and temporal distribution of interaction between groundwater and surface water (river, canal, lakes, spring flow).

(8) Locations of observation wells and Hydrograph of groundwater of groundwater head.

Step 2 Defining the flow system.

Conceptualize the movements of groundwater through the system.

(1) General direction of groundwater flow.

(2) Pattern of recharge and discharge.

(3) Connection between ground-surface water.

(4) Information for analysis—Groundwater head contour maps; Hydrochemical information; Isotopes; Groundwater temperature information; Hydrographs of groundwater head; Hydrographs of surface water level.

Step 3 Preparing the water budget.

Groundwater balance: Inflow+Outflow=Changes in storage.

Inflow—Precipitation; surface water; Underflow; Irrigation.

Outflow—Evapotranspiration; Spring flow; Baseflow to stream; Pumping; Underflow.

3. *Selecting computer code*

A computer code is a computer programme which solves the mathematical model of groundwater flow or contaminant transport numerically. There are many computer codes available. The selection of a suitable code depends on the complexity of the conceptual model and the purpose of study. The main considerations are:

(1) Types of model: flow model, particle tracking or solute transport model.

(2) Time dependency: steady or transient model.

(3) Dimensionality: one, two, quasi-three, or fully three dimensional model.

(4) Ability to describe the aquifer properties: homogeneous or heterogeneous; isotropic or anisotropic media.

(5) Ability to include various hydrological stresses.

(6) User friendliness.

(7) Requirements on the computer facility.

Widely-applied model—MODFLOW; MOC3D; MT3D; MODPATH; Processing Modflow (PM); Visual Modflfow (VM); Groundwater Modeling System (GMS); Finite Element Modflfow (FEM).

4. *Designing numerical model*

The design of numerical model includes the selection of modeling area, design of model grids, selection of stress periods and time steps, setting model boundaries and initial conditions. The conceptual model will be the bases for the design of the numerical model. The purpose of the modeling will dictate the sizes of grids and time steps. The memory and computing time of computers and the computer code may have limitations on total number of grids and time steps.

5. *Determination of model inputs*

The inputs to the model include initial and boundary conditions, hydrogeological parameters, and hydrological stresses. The data for all these inputs have to be entered to all grid points for all stress periods.

(1) Data for defining physical framework.

Geologic map and cross sections showing the areal and vertical extent and boundaries of the system.

1) Topographic map showing surface water bodies and divides.

2) Contour map of land surface elevation.

3) Contour maps showing the elevation of the base the stratigraphic units.

4) Maps showing the extent and thickness of stream and lake sediments.

(2) Hydrogeologic framework.

1) Schematization of aquifer systems.

2) Thickness and lateral extent of aquifers and confining beds.

3) Boundaries of aquifer systems.

4) Maps and cross sections showing the storage properties of the aquifers and confining beds.

5) Maps and cross sections showing the distribution of hydraulic conductivity/transmissivity.

6) Maps showing the extent and thickness of stream and lake sediments.

7) Groundwater head contour maps.

8) Locations of observation wells and measurements.

(3) Hydrological stresses.

1) Extent and rate of areal recharge (precipitation irrigation).

2) Extent and rate of areal discharge (evapotranspiration).

3) Locations and rate of wells (discharge/recharge).

4) Spatial and temporal distribution of interaction between groundwater and surface water (river, canal, lakes).

5) Spatial and temporal distribution of springs.

6) Locations of observation wells and hydrographs of groundwater head.

6. *Calibration of the model*

(1) Why to calibrate the model? The purpose of calibration is to establish the model that can reproduce the field measured groundwater heads or concentrations.

(2) How to calibrate the model? The calibration forces the model calculations approximate the field measured values through the adjustment of aquifer parameters or stresses by trial-and-error method or automated parameter estimation method requiring the measurements of groundwater heads or concentrations.

(3) Assessment of calibration. Mean error; maximum error; root mean square error (RMS).

(4) Sensitivity analysis. Objectives—Uncertainty of model parameters on model results; Identification of most important parameters.

Sensitivity coefficients—Head or concentration; RMS (root mean square).

Procedures for sensitivity analysis—Before and after model calibration; Systematic vary parameter values.

7. *Verification of the model*

To check whether the calibrated model has the predictive power, the calibrated model is applied to another period of time where a second field data are available. The model should also be able to reproduce the field measured values of groundwater heads or concentrations with hydrological stresses in this period.

8. *Application of the model*

The calibrated model is used to predict the response of the aquifer system to future events. In the prediction the model is run with calibrated aquifer parameters and future hydrological stresses. Some hydrological stresses are the proposed actions (such as abstraction). Others are natural uncontrolled stresses (such as recharge from precipitation).

9. *Presentation of results*

Clear presentation of modeling processing and results is essential for the effective communication of the modeling effort. The report on the modeling study should include chapters like:

(1) Introduction.

(2) Hydrogeological conceptual model.

(3) Numerical model setup.

(4) Model calibration.

(5) Model application.

(6) Summary and conclusions.

10. *Postaudit*

(1) Validation of the model prediction.

(2) Not yet a normal part of modeling.

(3) Groundwater models did not accurately predict the future due to ① error in conceptual model or/and ② errors in estimation of assumed future stresses.

Groundwater modeling is an iterative process. Steps outlined above may have to be repeated. Assumptions and even simplifications are necessary in the modeling because of the complexity of hydrogeological formations on the one hand and the lack of sufficient field data on the other hand. Models are only approximations of reality, but not reality itself. Therefore, groundwater modeling is not only a science but also an art. The science behind modeling can be learned relatively easily from many standard text books or short courses. However, the art of modeling can only be learned from practicing how to apply models. A successful modeler will have to know the science of the modeling and practice the art of modeling.

Great progress has been made in numerical modeling of groundwater in China since 1970's. Compared with the work done in some foreign countries, the numerical modeling of groundwater in China overlooked work steps of model purpose definition, sensitivity analysis, model postaudit and redesign. These steps play an important role in model validation and refinement. Groundwater model shouldn't be only constructed for answering pressing questions, it should be also developed for the purpose of groundwater system management and improved successively as new information source.

Chapter 2 Mathematical Model of Groundwater Flow

§ 2.1 Basic Concepts

1. *Underground water*

All water beneath the land surface is referred to as underground water (or subsurface water). The equivalent term for water on the land surface is surface water. Underground water occurs in two different zones. One zone, which occurs immediately below the land surface in most areas, contains both water and air and is referred to as the unsaturated zone. The unsaturated zone is almost invariably underlain by a zone in which all interconnected openings are full of water. This zone is referred to as the saturated zone.

Water in the saturated zone is the only underground water that is available to supply wells and springs and is the only water used reasonably. Recharge of the saturated zone occurs by percolation of water from the land surface through the unsaturated zone. The unsaturated zone is, therefore, of great importance to groundwater hydrology. This zone may be divided usefully into three parts: the soil zone, the intermediate zone, and the upper part of the capillary fringe.

The soil zone extends from the land surface to a maximum depth of a meter or two and is the zone that supports plant growth. It is crisscrossed by living roots, by voids left by decayed roots of earlier vegetation, and by animal and worm burrows. The porosity and permeability of this zone tend to be higher than those of the underlying material. The soil zone is underlain by the intermediate zone (Figure 2.1), which differs in thickness from place to place depending on the thickness of the soil zone and the depth to the capillary fringe.

The lowest part of the unsaturated zone is occupied by the capillary fringe, the sub zone between the unsaturated and saturated zones. The capillary fringe results from the attraction between water and rocks. As a result of this attraction, water clings as a film on the surface of rock particles and rises in small-diameter pores against the pull of gravity. Water in the capillary fringe and in the overlying part of the unsaturated zone is under a negative hydraulic pressure-that is, it is under a pressure less than the atmospheric (barometric) pressure. The water level is the level in the saturated zone at which the hydraulic pressure is equal to atmospheric pressure and is represented by the water level in unused wells. Below the water level, the hydraulic pressure increases with increasing depth.

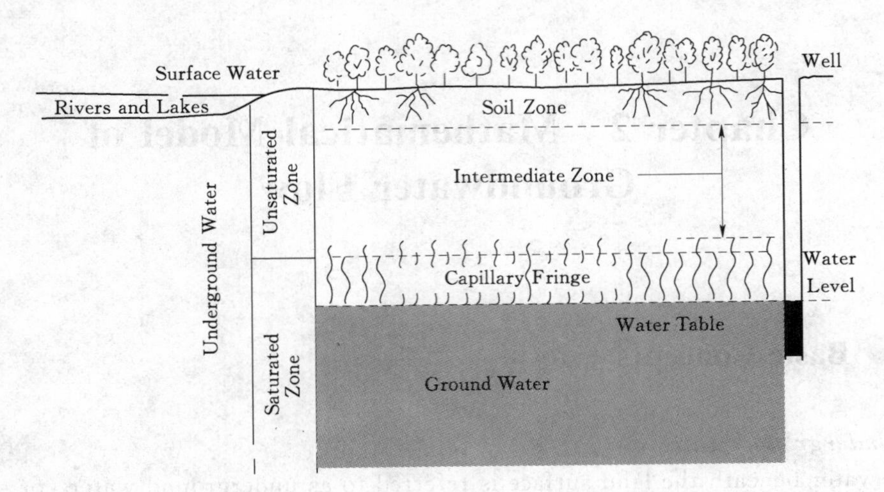

Figure 2. 1 Underground water

2. *Confining bed and aquifer*

From the standpoint of underground water occurrence, all rocks (includes unconsolidated sediments) that underlie the Earth's surface can be classified either as aquifers or as confining beds (Figure 2. 2).

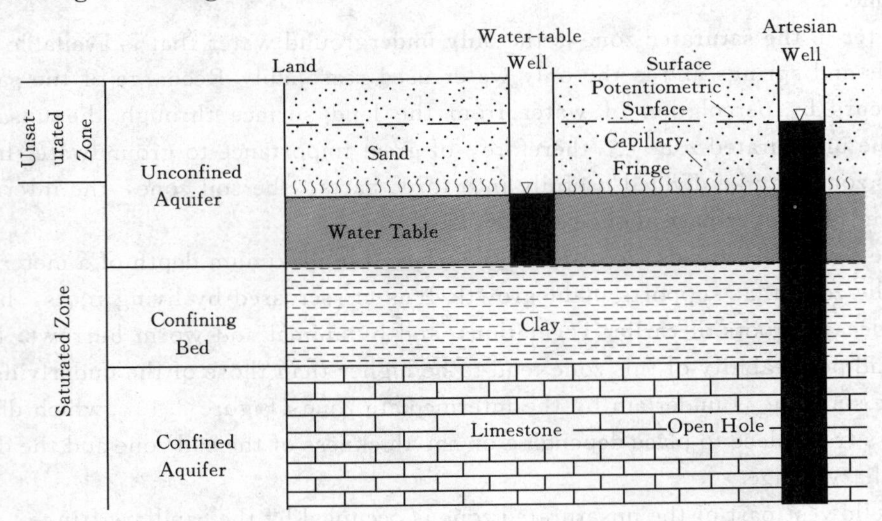

Figure 2. 2 Unconfined aquifer and confined aquifer

A confining bed is a rock unit having very low hydraulic conductivity that restricts the movement of ground water either into or out of adjacent aquifers.

Aquifer is a geologic unit that stores and transmits water, that is, an aquifer is a rock unit that will yield water in a usable quantity to a well or spring. Groundwater occurs in aquifers under two different conditions. Where water only partly fills an aquifer, the upper surface of the saturated zone is free to rise and decline. The water in such aquifers is said to be unconfined, and the aquifers are referred to as unconfined aquifers. Unconfined aquifers are also widely referred to as water-table aquifers. Where water completely fills an aquifer that is overlain by a confining bed, the water in the aquifer is said to be confined. Such aquifers are referred to as confined aquifers or as artesian aquifers.

Unconfined aquifer—Water is in contact with atmospheric pressure—Drill and well hit the water level. Although unconfined aquifers are used for water supply, they are often contaminated by wastes and chemicals at the surface.

Confined aquifer—Recharge up-gradient forces water to flow down and get trapped under an aquiclude. Water is under pressure due to the weight of the up-gradient water and the confinement of the water between "impermeable" layers. Water flows to surface under artesian pressure in an Artesian Well. Confined aquifers are less likely to be contaminated and thereby provide supplies of good quality.

3. *Porosity*

Porosity is the percentage volume occupied by voids and independent of scale, the ratio of openings (voids) to the total volume of a soil or rock is referred to as its porosity. Porosity is expressed either as a decimal fraction or as a percentage. Thus

$$n=\frac{V_v}{V_t}=\frac{V_t-V_s}{V_t} \tag{2.1}$$

Where: n is porosity as a decimal fraction; V_v is the volume of openings (voids); V_t is the total volume of a soil or rock sample; V_s is the volume of solids in the sample.

Porosity of unconsolidated deposits depends on the range in grain size (sorting) and on the shape of the rock particles but not on their size. Fine-grained materials tend to be better sorted and, thus, tend to have the largest porosities. For example, a pile of marbles and a pile of beach balls have spherical shape and differing sizes; the porosities are identical due to the similar shaping.

4. *Permeability*

Permeability measures the transmission property of the media and the interconnection of the pores. It is related to hydraulic conductivity and transmissivity. If the pore spaces between individual particles are connected, the water is capable of movement, both laterally and under gravity. Thus, although sandstone and mudstone are both quite porous, sandstone is permeable while mudstone is impermeable.

Good aquifer = High porosity+High permeability. For example, sand and gravel, sandstone, limestone, fractured rock, basalt.

Aqiuiclude, confining bed, aquitard—"Impermeable" unit forming a barrier to groundwater flow. For example, granite, shale, clay.

5. *Specific yield and specific retention*

Porosity tells us the maximum amount of water that a rock can contain when it is saturated. However, it is equally important to know that only a part of this water is available to supply a well or a spring. Hydrologists divide water in storage in the ground into the part that will drain under the influence of gravity (called specific yield) and the part that is retained as a film on rock surfaces and in very small openings (called specific retention). The physical forces that control specific retention are the same forces involved in the thickness and moisture content of the capillary fringe.

Specific yield tells how much water is available for man's use, and specific retention tells how much water remains in the rock after it is drained by gravity. Thus

$$\begin{cases} n = S_y + S_r \\ S_y = \dfrac{V_d}{V_t} \\ S_r = \dfrac{V_r}{V_t} \end{cases} \tag{2.2}$$

Where: n is porosity; S_y is specific yield; S_r is specific retention; V_d is the volume of water that drains from a total volume of V_t; V_r is the volume of water retained in a total volume of V_t; and V_t is total volume of a soil or rock sample.

In coarse sand the specific retention is low and the specific yield high. Fine clay has a high retention rate and a low specific yield, even though its porosity is quite high.

§ 2.2 Basic Law

2.2.1 Law of Energy Conservation

In 1856, a French engineer Henry Darcy conducted his famous experiment as sketched in Figure 2.3. A one-dimensional steady flow in the sand column was created by keeping water level constant in the left and right reservoirs.

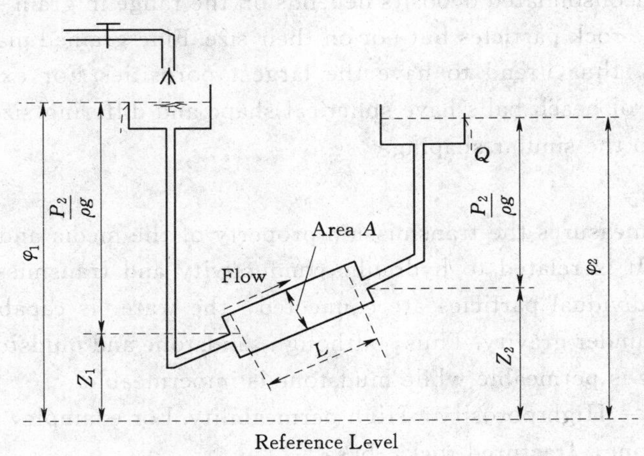

Figure 2.3 Darcy's experiment

Darcy measured the total discharge through a sand column when changing the difference between water levels in two reservoirs. He found the total discharge was proportional to the difference of water levels:

$$Q = KA\frac{\varphi_1 - \varphi_2}{L} = KAi \tag{2.3}$$

Equation (2.3) is the famous Darcy's law.

Where: Q is total discharge, $[L^3T^{-1}]$; A is cross-sectional area, $[L^2]$; K is hydraulic conductivity or coefficient of permeability, $[LT^{-1}]$; φ_1 and φ_2 are water levels in the left and right reservoirs, $[L]$, φ_1 is also the groundwater head of the left sand column, and φ_2 is the groundwater head of the right sand column; L is length of the sand column, [L]; i is hydraulic gradient.

1. *Groundwater head*

In general, groundwater head is defined as the elevation head plus the pressure head:

$$\varphi = z + \frac{p}{\rho g} \tag{2.4}$$

Where: z is elevation of the point concerned above the reference level, [L]; p is pressure in the fluid at that point, $[ML^{-1}T^{-2}]$; ρ is density of fluid (mass per unit volume), $[ML^{-3}]$; g is acceleration of gravity, $[LT^{-1}]$, $g=9.81\text{m/sec}^2$.

And the quantities z and $p/\rho g$ are usually called the elevation head and pressure head, respectively.

2. *Hydraulic gradient*

The rate of groundwater movement depends on the hydraulic gradient. The hydraulic gradient is the change in head per unit of distance in a given direction. If the direction is not specified, it is understood to be the direction in which the maximum rate of decrease of head occurs.

3. *Hydraulic conductivity and intrinsic permeability*

In Darcy's experiment, if using different fluids flowing through the same sand column, different values of the coefficient of permeability will be measured. It means that the coefficient of permeability does not only depend on the characteristics of the medium, but also on the properties of fluid. Hydraulic conductivity replaces the term "field coefficient of permeability" and should be used in referring to the water transmitting characteristic of material in quantitative terms. It is still common practice to refer in qualitative terms to "permeable" and "impermeable" material.

Hydraulic conductivity is not only different in different types of rocks but may also be different from place to place in the same rock. If the hydraulic conductivity is essentially the same in any area, the aquifer in that area is said to be homogeneous. If, on the other hand, the hydraulic conductivity differs from one part of the area to another, the aquifer is said to be heterogeneous. Hydraulic conductivity may also be different in different directions at any place in an aquifer. If the hydraulic conductivity is essentially the same in all directions, the aquifer is said to be isotropic. If it is different in different directions, the aquifer is said to be anisotropic.

Experiments show that the property of fluids influencing the hydraulic conductivity is the kinematic viscosity. The following relation is usually used:

$$K = \frac{k\rho g}{\mu} = \frac{kg}{\nu} \tag{2.5}$$

Where: k is called the intrinsic permeability with a dimension of, $[L^2]$, depending only on the characteristics of the medium; ρ is the mass density, $[ML^{-3}]$; μ is the dynamic viscosity, $[ML^{-1}T^{-1}]$, and measures the resistance of fluid to shearing that is necessary for flow; ν is kinematic viscosity, $[L^2T^{-1}]$, it is related to dynamic viscosity with $\nu = \frac{\mu}{\rho}$.

It is known that the kinematic viscosity varies with the temperature and density of fluids. In the cases flow of hot groundwater or salt groundwater, coefficient of permeability will be a function depending on the variation of the kinematic viscosity. However, the intrinsic permeability is a constant. Therefore, in these cases, the following alternative formulation of Darcy's law is more convenient:

$$q=-k\frac{g}{\nu}\frac{\mathrm{d}\varphi}{\mathrm{d}L} \tag{2.6}$$

For the case off fresh groundwater flow, the introduction of the intrinsic permeability has no advantages since the kinematic viscosity is a constant. Therefore, in the case of fresh groundwater flow, the coefficient of permeability is used.

The analog of laminar flow in tubes with groundwater flow in porous medium indicates that the intrinsic permeability is proportional to squared diameter of pore space (d^2) and porosity (n). In practice, empirical formula may be used to calculate the value of intrinsic permeability. One of such formula is Kozeny—Carman's formula:

$$k=cd^2\frac{n^3}{(1-n)^2} \tag{2.7}$$

Where: d is diameter of pore space, [L]; n is porosity; c is constant to be determined with the experiment.

Equation (2.7) explains why the permeability of clay is small and gravel is large. Although the porosity of clay is very large, the pore space is very small resulting of low permeability. On contrary, the pore space of gravel is very large and therefore, the permeability of gravel is high. Table 2.1 gives range of permeability for common porous medium. In reality the structure of porous medium is so complicated that no direct formula can be used to calculate the permeability. Its value is usually determined with pumping tests.

Table 2.1 Range of permeability for common porous medium

Media	k (m^2)	K (m/d)
Clay	$10^{-17}\sim10^{-15}$	$10^{-5}\sim10^{-3}$
Slit	$10^{-15}\sim10^{-13}$	$10^{-3}\sim10^{-1}$
Sand	$10^{-12}\sim10^{-10}$	$10^{-1}\sim10^{2}$
Gravel	$10^{-9}\sim10^{-8}$	$10^{4}\sim10^{2}$

4. *Specific discharge*

The specific discharge is defined as the discharge per unit of cross-sectional area and denoted by q. It follows from equation (2.3) that,

$$q=K\frac{\Delta\varphi}{\Delta l} \tag{2.8}$$

By taking the limit $\Delta l\rightarrow 0$, the equation (2.8) becomes,

$$q=-K\frac{\mathrm{d}\varphi}{\mathrm{d}L} \tag{2.9}$$

Where: q is specific discharge, [LT^{-1}]; $-\mathrm{d}\varphi/\mathrm{d}L$ is gradient of groundwater head (hydraulic gradient).

Equation (2. 9) is the differential formulation of the Darcy's law. It expresses a linear relationship between the specific discharge and hydraulic gradient.

5. *Velocity*

Although the specific discharge has the same dimension of the velocity, it is not the actual velocity of the groundwater flow. The total cross-sectional area is A, but the area through which the groundwater can flow is only nA. n is the porosity of the sand. Hence, in Darcy's experiment, the actual velocity (v) of the flow can be computed as:

$$v=\frac{Q}{nA}=\frac{q}{n}=\frac{\overline{v}}{n} \tag{2.10}$$

Where: v is actual velocity, $[LT^{-1}]$; $\overline{v}$ is seepage velocity, $[LT^{-1}]$.

The actual velocity is always greater than the seepage velocity since the porosity is smaller than one.

6. *Validity of Darcy's law*

The linear relationship between the specific discharge and hydraulic gradient suggests that the Darcy's law can be applied to the fluid flow. Therefore, Renolds number can be used as an indicator of validity of the Darcy's law. The Renolds number for porous medium is defined as:

$$Re=\frac{qd}{\nu}=\frac{\rho\overline{v}d}{\mu}=\frac{Inertial\quad forces}{Viscous\quad forces} \tag{2.11}$$

Where: d is average grain diameter, $[L]$; ν is kinematic viscosity, $[L^2T^{-1}]$.

Experiments show that when $Re<10$, flow is laminar fluid flow and Darcy's law can be applied. Practical experiences show that Darcy's law can be applied to most cases of groundwater flow in porous medium.

Darcy's law in the differential form of equation (2. 9) expresses the relationship between the specific discharge and hydraulic gradient for the uniform flow in one dimension. Darcy's law can be extended to more general cases of groundwater flow.

For three-dimensional groundwater flow in isotropic porous medium, the hydraulic conductivity (K) having the same value in all directions and is a scalar. Darcy's law will be written as:

$$\xrightarrow{\text{Gradient}} K \xrightarrow{\text{Flow}}$$

$$\begin{cases} q_x=-K\dfrac{\partial\varphi}{\partial x} \\ q_y=-K\dfrac{\partial\varphi}{\partial y} \\ q_z=-K\dfrac{\partial\varphi}{\partial z} \end{cases} \tag{2.12}$$

Where: q_x, q_y and q_z are three flow components in x, y and z directions in isotropy medium respectively.

The groundwater head will be a function of x, y and z coordinates and is defined as:

$$\varphi(x,y,z)=z+\frac{p(x,y,z)}{\rho g} \tag{2.13}$$

Where: ρ is assumed a constant.

In cases of three-dimensional groundwater flow in anisotropic medium, the hydraulic conductivity (K) having directional properties and is really a tensor in 3D. Darcy's law will be written as:

Gradient → K → Flow

$$\begin{cases} q_x = -K_{xx}\dfrac{\partial\varphi}{\partial x} - K_{xy}\dfrac{\partial\varphi}{\partial y} - K_{xz}\dfrac{\partial\varphi}{\partial z} \\ q_y = -K_{yx}\dfrac{\partial\varphi}{\partial x} - K_{yy}\dfrac{\partial\varphi}{\partial y} - K_{yz}\dfrac{\partial\varphi}{\partial z} \\ q_z = -K_{zx}\dfrac{\partial\varphi}{\partial x} - K_{zy}\dfrac{\partial\varphi}{\partial y} - K_{zz}\dfrac{\partial\varphi}{\partial z} \end{cases} \tag{2.14}$$

The conductivity ellipse (anisotropic vs. isotropic) shown as Figure 2.4.

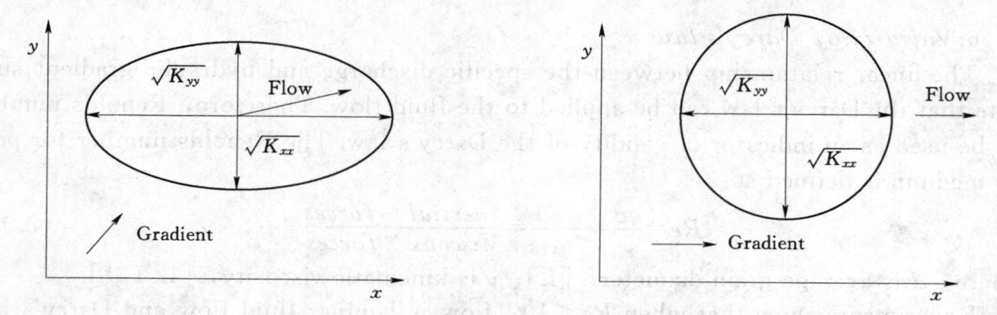

Figure 2.4 Conductivity ellipse (anisotropic vs. isotropic)

If $K_{yy} = K_{xx}$ then the media is isotropic and ellipse is a circle. It is convenient to describe Darcy's law as:

$$\vec{q} = -\overline{\overline{K}}\,\nabla\varphi \tag{2.15}$$

Where: ∇ is called del and is a gradient operator, so $\nabla\varphi$ is the gradient in all three directions (in 3D), $\nabla\varphi = \begin{bmatrix} \dfrac{\partial\varphi}{\partial x} \\ \dfrac{\partial\varphi}{\partial y} \\ \dfrac{\partial\varphi}{\partial z} \end{bmatrix}$; $\overline{\overline{K}} = \begin{bmatrix} K_{xx} & K_{xy} & K_{xz} \\ K_{yx} & K_{yy} & K_{yz} \\ K_{zx} & K_{zy} & K_{zz} \end{bmatrix}$.

The magnitudes of K in the principal directions are known as the principal conductivities. If the coordinate axes are aligned with the principal directions of the conductivity tensor, then Darcy's law can be generalized as:

$$\begin{cases} q_x = -K_{xx}\dfrac{\partial\varphi}{\partial x} \\ q_y = -K_{yy}\dfrac{\partial\varphi}{\partial y} \\ q_z = -K_{zz}\dfrac{\partial\varphi}{\partial z} \end{cases} \tag{2.16}$$

Where: K_{xx}, K_{yy} and K_{zz} are anisotropic hydraulic conductivity in three principal directions, respectively.

2.2.2 Law of Mass Conservation

Groundwater flow also satisfies the principle of mass conservation, or mass balance. Taking a small parallelepiped in porons medium, groundwater flow through the parallelepiped must obey the mass balance stated as:

Total mass in－total mass out ＝ change of mass storage

If the density of groundwater is a constant, mass balance will be identical to water balance. Water balance states that

Total flow in－total flow out ＝ change of water storage

From the principle of mass balance, the so-called equation of continuity can be derived.

Figure 2.5 shows the mass fluxes through six sides of a parallelepiped in porous medium.

First, consider flow in x-direction only and q_x $[L^3/L^2-T]$ is specific discharge in x-direction (volume flux per area) at a point (x, y, z). The mass flow through plane $y-z$ at (x, y, z) is $\rho q_x \mathrm{d}y\mathrm{d}z$ [M/T]. Figure 2.6 shows the mass fluxes through 6 sides of a parallelepiped in porous medium.

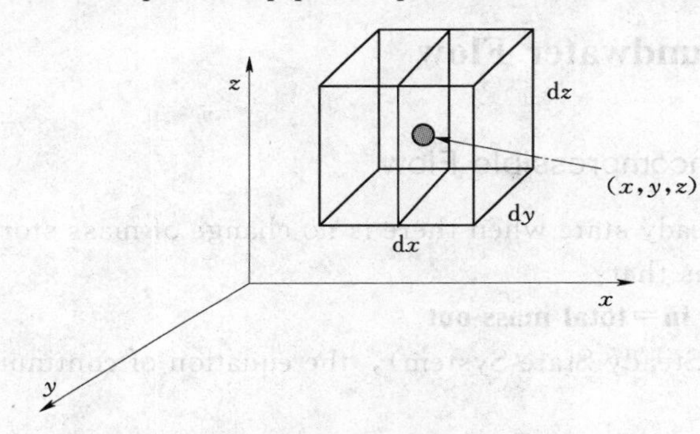

Figure 2.5 Rectangular parallelepipeds in porous medium

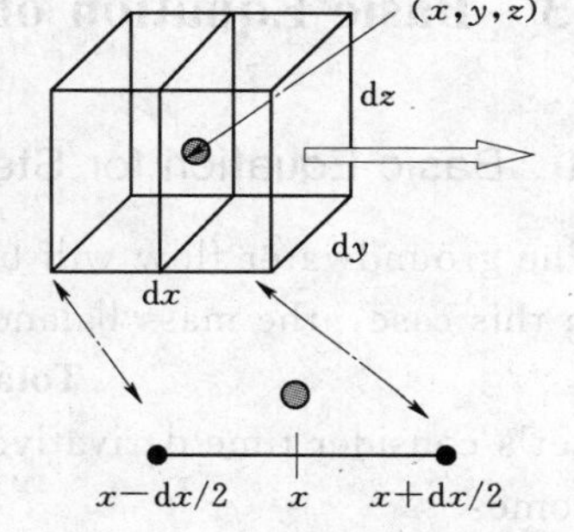

Figure 2.6 Mass fluxes through 6 sides of a parallelepiped

Rate of change of mass flux in the x-direction per unit time per cross-section is $\frac{\partial}{\partial x}[\rho q_x]\mathrm{d}y\mathrm{d}z$.

Mass flow into the entry plane $y-z$ is:

$$[\rho q_x]\mathrm{d}y\mathrm{d}z-\frac{\partial}{\partial x}[\rho q_x]\frac{\mathrm{d}x}{2}\mathrm{d}y\mathrm{d}z$$

And mass flow out of the exit plane $y-z$ is:

$$[\rho q_x]\mathrm{d}y\mathrm{d}z+\frac{\partial}{\partial x}[\rho q_x]\frac{\mathrm{d}x}{2}\mathrm{d}y\mathrm{d}z$$

In the x-direction, the flow in minus the flow out is:

$$-\frac{\partial}{\partial x}[\rho q_x]\mathrm{d}x\mathrm{d}y\mathrm{d}z \tag{2.17}$$

Similarly, the flow in the y-direction through the plane $\mathrm{d}x\mathrm{d}z$ can be derived as:

$$-\frac{\partial}{\partial y}[\rho q_y]\mathrm{d}x\mathrm{d}y\mathrm{d}z \tag{2.18}$$

Similarly, we get for the net mass flux in the z-direction:

$$-\frac{\partial}{\partial z}[\rho q_z]\mathrm{d}x\mathrm{d}y\mathrm{d}z \tag{2.19}$$

The total mass in the parallelepiped $\mathrm{d}x\mathrm{d}y\mathrm{d}z$ is:

$$M=\rho(\mathrm{d}x\mathrm{d}y\mathrm{d}z)n$$

The change (increase) of mass storage per unit time in the parallelepiped is:

$$\frac{\partial M}{\partial t}=\frac{\partial}{\partial t}[\rho(\mathrm{d}x\mathrm{d}y\mathrm{d}z)n] \tag{2.20}$$

According to the mass balance, the change of mass storage must be equal to total increase of mass due to flow in x, y, and z directions. Thus, the summation of equations (2.17), (2.18) and (2.19) equals equation (2.20) as:

$$\frac{\partial}{\partial t}[\rho(\mathrm{d}x\mathrm{d}y\mathrm{d}z)n]=-\left[\frac{\partial(\rho q_x)}{\partial x}+\frac{\partial(\rho q_y)}{\partial y}+\frac{\partial(\rho q_z)}{\partial z}\right]\mathrm{d}x\mathrm{d}y\mathrm{d}z \tag{2.21}$$

Equation (2.21) is the general equation of continuity of groundwater flow. The equation of continuity is second fundamental equation of groundwater flow. The equation of continuity combined with Darcy's law will result in basic equations describing groundwater flow in porous medium.

§ 2.3 Basic Equation of Groundwater Flow

2.3.1 Basic Equation for Steady Incompressible Flow

The groundwater flow will be in steady state when there is no change of mass storage. In this case, the mass balance states that:

Total mass in = total mass out

Let's consider time derivative=0 (Steady State System), the equation of continuity become:

$$\frac{\partial(\rho q_x)}{\partial x}+\frac{\partial(\rho q_y)}{\partial y}+\frac{\partial(\rho q_z)}{\partial z}=0 \tag{2.22}$$

Furthermore, if the fluid is incompressible, the density ρ is a constant and equation (2.22) reduces to:

$$\frac{\partial q_x}{\partial x}+\frac{\partial q_y}{\partial y}+\frac{\partial q_z}{\partial z}=0 \tag{2.23}$$

Equation (2.23) is the equation of continuity for steady incompressible groundwater flow.

Substitution of Darcy's law, equation (2.12) and equation (2.23) gives:

$$\frac{\partial^2\varphi}{\partial x^2}+\frac{\partial^2\varphi}{\partial y^2}+\frac{\partial^2\varphi}{\partial z^2}=0 \tag{2.24}$$

This is the basic differential equation of steady incompressible flow in homogeneous isotropic porous medium. It is noted that equation (2.24) is the standard **Laplace equation** and is often written in abbreviation form:

$$\nabla^2\varphi=0 \tag{2.25}$$

Substitution of Darcy's law, equation (2.16) into equation (2.23) will give the differential equation of steady incompressible flow in anisotropic porous medium, which is:

$$\frac{\partial}{\partial x}\left(K_{xx}\frac{\partial \varphi}{\partial x}\right)+\frac{\partial}{\partial y}\left(K_{yy}\frac{\partial \varphi}{\partial y}\right)+\frac{\partial}{\partial z}\left(K_{zz}\frac{\partial \varphi}{\partial z}\right)=0 \tag{2.26}$$

When the porous medium is anisotropic but homogeneous, equation (2.26) reduces to:

$$K_{xx}\frac{\partial^2 \varphi}{\partial x^2}+K_{yy}\frac{\partial^2 \varphi}{\partial y^2}+K_{zz}\frac{\partial^2 \varphi}{\partial z^2}=0 \tag{2.27}$$

2.3.2 Basic Equation for Non-steady Compressible Flow

In the case of non-steady flow, the storage of mass is changed since the total mass in is not equal to the total mass out.

From the principles of Soil Mechanics, it is known that the total pressure (σ_z) from overlying geological strata is balanced by the effective pressure (σ_z') of grains and pore pressure (p) of water (Figure 2.7). i.e..

$$\sigma_z=\sigma_z'+p \tag{2.28}$$

Since the total pressure can be assumed to be independent of time, any increase of pore pressure will result in a decrease of effective pressure, the consequences of such a change are:

(1) Water is compressed due to the increase of pore pressure. Therefore, the density of water is a function of time.

(2) Grain skeleton is expanded due to the decrease of effective pressure. Hence, the porosity and size of grain skeleton are functions of time.

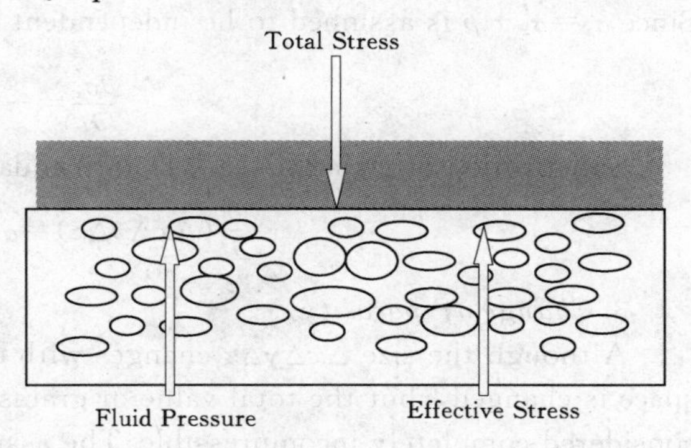

Figure 2.7 Principle of effective stress

Taking the above effects into consideration, the change of mass storage in equation (2.20) can be expressed as:

$$\frac{\partial}{\partial t}[\rho(\Delta x\Delta y\Delta z)n]=(\Delta x\Delta y\Delta z)n\frac{\partial \rho}{\partial t}+\rho(\Delta x\Delta y\Delta z)\frac{\partial n}{\partial t}+\rho n\frac{\partial}{\partial t}(\Delta x\Delta y\Delta z) \tag{2.29}$$

1. *Compressibility of water*

From the elastic theory, the relative change of density of water is proportional to the change of pore pressure, i.e..

$$\frac{\mathrm{d}\rho}{\rho}=\beta_w \mathrm{d}p \tag{2.30}$$

Where: β_w is the coefficient of compressibility of water. Its value is about 0.5×10^{-9} m^2/N.

From equation (2.30) it follows that:

$$\frac{\partial \rho}{\partial t}=\frac{\mathrm{d}\rho}{\mathrm{d}p}\frac{\partial p}{\partial t}=\beta_w\rho\frac{\partial p}{\partial t} \tag{2.31}$$

2. *Compressibility of soil*

From the theory of elasticity, soil will be compressed under the pressure. When sample is confined horizontally, the vertical relative compression is proportional to the increase of pressure, i. e. .

$$\frac{d\Delta z}{\Delta z}=-\alpha d\sigma_z' \tag{2.32}$$

Where: σ_z' is vertical pressure on sol with area $(\Delta x\Delta y)$ and height Δz; α is coefficient of compressibility of soil, common values are $10^{-8}\sim10^{-7}$ m²/N for sand, and $10^{-7}\sim10^{-6}$ m²/N for clay.

In many practical cases usually the horizontal deformations of the soil skeleton is much smaller than the vertical deformation. Therefore, it can be assumed that Δx, Δy are constant and only Δz changes. Under this assumption,

$$\frac{\partial}{\partial t}(\Delta x\Delta y\Delta z)=\frac{\partial(\Delta z)}{\partial t}\Delta x\Delta y=\frac{\partial(\Delta z)}{\partial\sigma_z'}\frac{\partial\sigma_z'}{\partial t}\Delta x\Delta y=-\alpha\Delta x\Delta y\Delta z\frac{\partial\sigma_z'}{\partial t} \tag{2.33}$$

Since $\sigma_z=\sigma_z'+p$ is assumed to be independent of time, i. e. $\frac{\partial\sigma_z}{\partial t}=0$, so that:

$$\frac{\partial\sigma_z'}{\partial t}=-\frac{\partial p}{\partial t} \tag{2.34}$$

Substitution of equation (2. 34) into equation (2. 33) yields:

$$\frac{\partial}{\partial t}(\Delta x\Delta y\Delta z)=\alpha\Delta x\Delta y\Delta z\frac{\partial p}{\partial t} \tag{2.35}$$

3. *Change of porosity*

Although the size $\Delta x\Delta y\Delta z$ changes with compression, it is assumed that only pore space is changed, but the total value of grains (V_s) is kept constant since grains can be considered completely incompressible. The assumption of $V_s=(1-n)\Delta x\Delta y\Delta z$ constant gives:

$$\frac{\partial V_s}{\partial t}=\frac{\partial(1-n)}{\partial t}\Delta x\Delta y\Delta z+(1-n)\Delta x\Delta y\frac{\partial(\Delta z)}{\partial t}=0 \tag{2.36}$$

From equation (2. 36) with equation (2. 35), the relation between changes of porosity with change pore pressure is found as:

$$\frac{\partial n}{\partial t}=\frac{1-n}{\Delta z}\frac{\partial(\Delta z)}{\partial t}=(1-n)\alpha\frac{\partial p}{\partial t} \tag{2.37}$$

4. *Equation of continuity of non-steady compressible flow*

Substitution of equations (2. 31), (2. 35) and (2. 37) into equation (2. 29) gives:

$$\frac{\partial}{\partial t}[\rho(\Delta x\Delta y\Delta z)n]=\Delta x\Delta y\Delta z\rho(\alpha+n\beta_w)\frac{\partial p}{\partial t} \tag{2.38}$$

Substitution of equation (2. 38) into equation (2. 21) gives the general equation of continuity for non-steady compressible flow as:

$$\rho(\alpha+n\beta_w)\frac{\partial p}{\partial t}=-\left[\frac{\partial(\rho q_x)}{\partial x}+\frac{\partial(\rho q_y)}{\partial y}+\frac{\partial(\rho q_z)}{\partial z}\right] \tag{2.39}$$

5. *Basic equation of non-steady compressible flow*

In case of compressible flow in homogeneous isotropic porous medium, Darcy's law of equation (2. 12) will be replaced by:

$$\begin{cases} q_x = -\dfrac{K}{\rho g}\dfrac{\partial p}{\partial x} \\ q_y = -\dfrac{K}{\rho g}\dfrac{\partial p}{\partial y} \\ q_z = -K - \dfrac{K}{\rho g}\dfrac{\partial p}{\partial z} \end{cases} \tag{2.40}$$

Substitution of equation (2.40) into equation (2.39) results in:

$$\rho g(\alpha + n\beta_w)\frac{\partial p}{\partial t} = K\left(\frac{\partial^2 p}{\partial x^2} + \frac{\partial^2 p}{\partial y^2} + \frac{\partial^2 p}{\partial z^2}\right) + g\frac{\partial \rho}{\partial z} \tag{2.41}$$

If the vertical gradient of fluid density can be neglected (for example in one fluid flow), equation (2.41) reduces to:

$$\frac{\partial^2 p}{\partial x^2} + \frac{\partial^2 p}{\partial y^2} + \frac{\partial^2 p}{\partial z^2} = \frac{\rho g(\alpha + n\beta_w)}{K}\frac{\partial p}{\partial t} \tag{2.42}$$

This is a basic equation of non-steady compressible flow in homogeneous isotropic medium.

Using the equation of $\varphi = z + \dfrac{p}{\rho g}$ an equation in terms of head can be found as:

$$\frac{\partial^2 \varphi}{\partial x^2} + \frac{\partial^2 \varphi}{\partial y^2} + \frac{\partial^2 \varphi}{\partial z^2} = \frac{S_s}{K}\frac{\partial \varphi}{\partial t} \tag{2.43}$$

Where: $S_s = \rho g + (\alpha + n\beta_w)$ is usually called the storativity or specific storage coefficient. It is defined as the volume of water stored in a unit volume of soil by a unit increase of the head. It has a dimension of $[L^{-1}]$. The physical explanation of the storativity is that with increase of head more water is stored in pore space since water is compressed; and pore space is enlarged since the grain skeleton is expanded. Since the expansion and compression of water and soil skeleton is elastic, S_s is called elastic storage.

§ 2.4 Boundary Conditions

2.4.1 Boundary Conditions

Boundary conditions are mathematical statements specifying the groundwater heads or flux of groundwater flow at the boundaries of the model domain. Physical boundaries of groundwater flow systems are formed by the physical presence of an impermeable body of rock or a large body of surface water. Other boundaries form as a result of hydrologic conditions. These invisible boundaries are called hydraulic boundaries which include groundwater divides and streamlines. Figure 2.8 shows a regional flow system bounded by physical boundaries:

(1) Impermeable rock at the bottom.

(2) Water level at the top.

(3) Rivers on the left and the right sides.

Groundwater divided in the middle separates the flow system into two sub-flow systems.

Hydrogeological boundaries are represented by three of mathematical conditions.

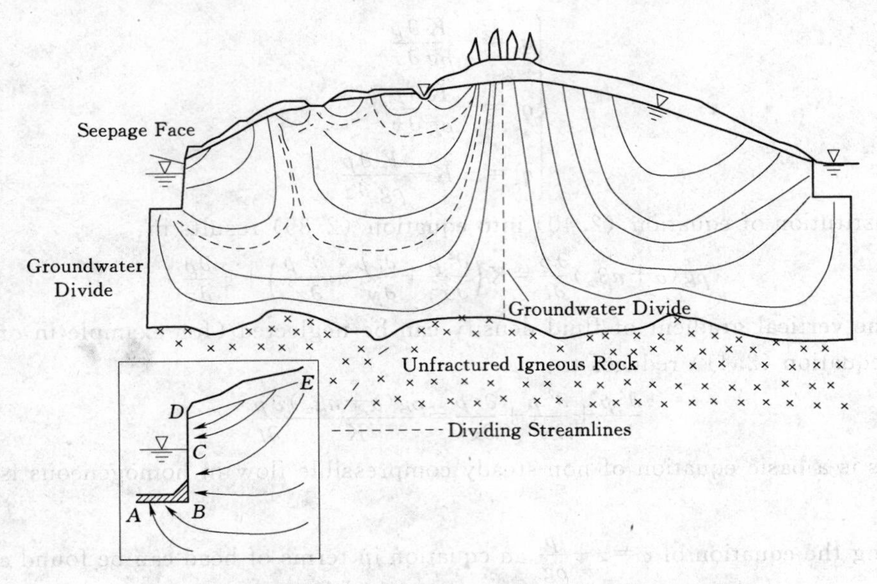

Figure 2.8 Regional flow systems showing boundary conditions

First type is called specified head boundaries (Dirichlet conditions). Along these boundaries, groundwater heads are known and specified as:

$$\varphi|_{\Gamma_1}=\varphi(x,y,z) \quad (x,y,z)\in\Gamma_1 \tag{2.44}$$

Where: Γ_1 represents locations of the first type boundaries. River in Figure 2.8 can be treated as the specified head boundaries.

Second type is called specified flow boundaries (Neumann conditions). Across these boundaries groundwater flow is known and specified as:

$$K\frac{\partial\varphi}{\partial n}|_{\Gamma_2}=q(x,y,z) \quad (x,y,z)\in\Gamma_2 \tag{2.45}$$

Where: Γ_2 represents locations of the second type boundaries. A special case of specified flow boundary is no-flow boundary where flux is specified to be zero. Groundwater divide in Figure 2.8 can be treated as a no-flow boundary.

Third type is called head-dependant flow boundaries (Cauchy or mixed boundary conditions). Groundwater flow across these boundaries depends on the head at boundaries. These boundaries are specified as:

$$q(x,y,z)|_{\Gamma_3}=K'\frac{\varphi-\varphi_0}{B'} \quad (x,y,z)\in\Gamma_3 \tag{2.46}$$

Where: Γ_3 represents locations of the third type boundaries. Rivers in Figure 2.8 can be also treated as head-dependant flow boundaries if the river bed deposits are different from aquifer medium.

2.4.2 Initial Conditions

Initial conditions refer to the distribution of groundwater heads everywhere in the aquifer at the beginning of a reference time. It is necessary only for transient groundwater flow. The initial condition is specified as:

$$\varphi|_{t=0}=\varphi_0(x,y,z) \tag{2.47}$$

§ 2. 5 Mathematical Model

Mathematical model of groundwater flow consist of a governing equation (partial differential equations of groundwater flow), boundary conditions, and initial conditions.

Mathematical model of the steady state groundwater flow consists of a governing equation of steady flow [equation (2. 24)] and boundary conditions only. A unique solution of the model exits only if one of boundary conditions is head specified boundary. An example of steady flow model is given as:

$$\frac{\partial^2 \varphi}{\partial x^2}+\frac{\partial^2 \varphi}{\partial z^2}=0 \tag{2. 48a}$$

$$\varphi|_{\Gamma_1}=\varphi_1(x) \quad x\in\Gamma_1 \tag{2. 48b}$$

$$K\frac{\partial \varphi}{\partial z}|_{\Gamma_2}=0 \quad z\in\Gamma_2 \tag{2. 48c}$$

Mathematical model equation (2. 48) describes steady groundwater flow in the left subsystem of Figure 2. 8. The governing equation (2. 48a) describes two-dimensional cross sectional steady state flow. Specified head boundaries equation (2. 48b) include water level at the top and the river at the left side. No-flow boundaries equation (2. 48c) include groundwater divide along the left side boundary and the right side divide separating the flow system.

Chapter 3 Numerical Method—Finite-Difference Method

§ 3. 1 Main Ideas and Solving Steps

3. 1. 1 Main Ideas of Finite-Difference Method

A numerical solution which transforms differential equation problems into difference equation problems is called finite difference method in mechanics. The basic idea of the finite difference method is replacing a continuous definite area by grids constituted by finite discrete points (called grid nodes). Using the functions of discrete variables which defined on the grids to approximate the functions of continuous variables on a continuous definite area; approximating the derivative and integral by difference quotient and integral sum respectively in original equation and Boundary conditions, so we can use algebraic equations, that is finite difference equations, approximately substitute original differential equations and Boundary conditions. The solutions of this equation are the approximate solutions in discrete points of the original equation. Then, we could obtain the approximate solutions of definite problem from discrete solution in the whole area with the interpolation method.

When we solving partial differential equation with numerical methods, the partial differential equation is transformed into algebraic equation if every derivatives are approximately substituted by finite difference formula, which is definite difference method.

The steps to solve partial differential equation by definite difference method are shown as following:

(1) Zone discretization, namely the given computing region of partial differential equation is subdivided into grids composed by the finite mesh points.

(2) Approximate substitution, namely substituting the derivative of each point with the finite difference formula.

(3) Solved by approximation. In other words, this process can be viewed as using interpolation polynomial and its differential to replace partial differential equation.

3. 1. 2 Steps of Solving Groundwater Flow Problem by Difference Method

The main steps solving the groundwater flow problems by difference method are shown as following:

(1) Subdivide the vadose region and determine the discrete points. The studied vadose zone is divided into a grid system by some sort of geometric shapes (such as rectan-

gles, arbitrary polygon). For a rectangular grid, the distance in the x-direction (Δx) and y-direction distance (Δy) is usually known as the space step. After the subdivision of vadose region, we can determine the discrete points, usually referred to as spatial discretization. Discrete points lies in the intersection of the grid (called nodes) or grid center (this point is called the grid points). In essence, solving the groundwater flow problems by numerical method is to calculate the value of discrete point in seepage zone. Moreover, for unstable flow problems, we should discrete the time also, that is dividing the continuous period into equal or unequal period Δt_i, which is usually called time step.

(2) Establish the Differential equations for groundwater flow problems. According to Darcy's law and the principle of water balance, establishing mathematical models that describe the groundwater flow problem, and then use the difference quotient instead of the derivative, thus the solving of a definite problem is transformed into a difference equation.

(3) Solving the algebraic (differential) equation set.

§ 3.2 Finite Difference Formulae

3.2.1 Finite Difference Approximation

The basic principle of finite difference method is to approximately express the derivative of the function of the hydraulic head at a certain node with the heads of the node and its neighboring nodes and the distances between them. Actually, finite difference equation is the approximate expression of basic partial differential equations, and the approximate degree can be studied by the Taylor series.

Take a node x on the x-axis as shown in Figure 3.1, the coordinates at the left and right of the node with a distance of Δx are $x-\Delta x$ and $x+\Delta x$ respectively.

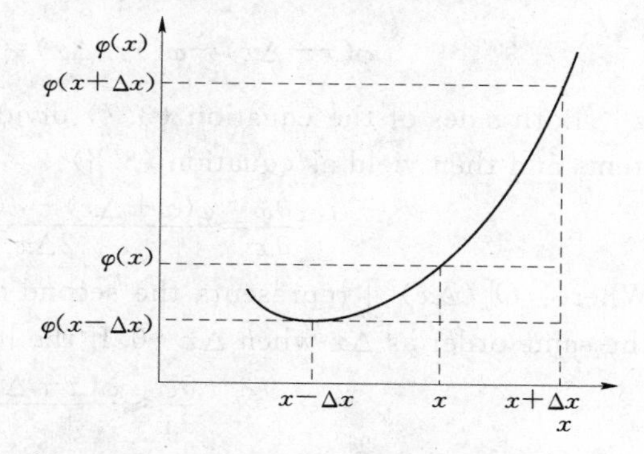

Figure 3.1 Finite difference approximation with first-order derivative

Taking the node x as the center, developing the function of heads $\varphi(x)$ by Taylor series along the positive and negative x direction.

$$\varphi(x+\Delta x)=\varphi(x)+\Delta x\frac{\partial\varphi}{\partial x}+\frac{(\Delta x)^2}{2!}\frac{\partial^2\varphi}{\partial x^2}+\frac{(\Delta x)^3}{3!}\frac{\partial^3\varphi}{\partial x^3}+\frac{(\Delta x)^4}{4!}\frac{\partial^4\varphi}{\partial x^4}+\cdots \tag{3.1}$$

$$\varphi(x-\Delta x)=\varphi(x)-\Delta x\frac{\partial\varphi}{\partial x}+\frac{(\Delta x)^2}{2!}\frac{\partial^2\varphi}{\partial x^2}-\frac{(\Delta x)^3}{3!}\frac{\partial^3\varphi}{\partial x^3}+\frac{(\Delta x)^4}{4!}\frac{\partial^4\varphi}{\partial x^4}-\cdots \tag{3.2}$$

Where: $\varphi(x)$ represents the head at node x; $\frac{\partial\varphi}{\partial x}$ represents the derivative value of the head at node x.

To transpose the terms of equation (3.1), then divide by Δx, we get:

$$\frac{\varphi(x+\Delta x)-\varphi(x)}{\Delta x}=\frac{\partial\varphi}{\partial x}+\frac{\Delta x}{2!}\frac{\partial^2\varphi}{\partial x^2}+\frac{(\Delta x)^2}{3!}\frac{\partial^3\varphi}{\partial x^3}+\cdots$$

Or be adapted as

$$\frac{\partial\varphi}{\partial x}=\frac{\varphi(x+\Delta x)-\varphi(x)}{\Delta x}+O(\Delta x) \tag{3.3}$$

Where: $O(\Delta x)$ represents first-order truncation error, is an infinitesimal of the same order as Δx when $\Delta x\to 0$. If we omit the item $O(\Delta x)$, the finite difference approximation with the first-order derivative will be expressed as equation (3.4):

$$\frac{\partial\varphi}{\partial x}\approx\frac{\varphi(x+\Delta x)-\varphi(x)}{\Delta x} \tag{3.4}$$

Equation (3.4) is the forward-difference formula with first-order derivative, having a first-order truncation error.

Similarly, by equation (3.2), we can get:

$$\frac{\partial\varphi}{\partial x}=\frac{\varphi(x)-\varphi(x-\Delta x)}{\Delta x}+O(\Delta x) \tag{3.5}$$

To omit the item $O(\Delta x)$, we can get:

$$\frac{\partial\varphi}{\partial x}\approx\frac{\varphi(x)-\varphi(x-\Delta x)}{\Delta x} \tag{3.6}$$

Equation (3.6) is the backward-difference formula of first-order derivative, the first-order truncation error is $O(\Delta x)$.

If the equation (3.1) minus equation (3.2), we can get:

$$\varphi(x+\Delta x)-\varphi(x-\Delta x)=2\Delta x\frac{\partial\varphi}{\partial x}+2\frac{(\Delta x)^3}{3!}\frac{\partial^3\varphi}{\partial x^3}+\cdots \tag{3.7}$$

Both sides of the equation (3.7) divide by 2 Δx at the same time, arranging the items and then yield as equation (3.8):

$$\frac{\partial\varphi}{\partial x}=\frac{\varphi(x+\Delta x)-\varphi(x-\Delta x)}{2\Delta x}+O[(\Delta x)^2] \tag{3.8}$$

Where: $O[(\Delta x)^2]$ represents the second-order truncation error, is an infinitesimal of the same order as Δx when $\Delta x\to 0$. If the item $O[(\Delta x)^2]$ is omitted, we can get:

$$\frac{\partial\varphi}{\partial x}\approx\frac{\varphi(x+\Delta x)-\varphi(x-\Delta x)}{2\Delta x} \tag{3.9}$$

Equation (3.9) is the central-difference formula of first-order derivative, having a second-order truncation error.

From three types of finite difference formula with first-order derivative mentioned as above, we can find out that one-side difference formula (forward-difference and backward-difference) has a first-order truncation error while central-difference formula has second-order truncation error. Therefore, central-difference formula is more accurate than one-side difference formula.

In order to get the finite difference expression of second-order derivative for the function of heads $\varphi(x)$ at node x, adding equation (3.1) and equation (3.2) then get:

$$\varphi(x+\Delta x)+\varphi(x-\Delta x)=2\varphi(x)+2\frac{(\Delta x)^2}{2!}\frac{\partial\varphi^2}{\partial x^2}+O[(\Delta x)^4] \tag{3.10}$$

Both sides of the equation (3.10) divide by $(\Delta x)^2$ at the same time, arranging the items and then yield as equation (3.11):

$$\frac{\partial^2 \varphi}{\partial x^2}=\frac{\varphi(x+\Delta x)-2\varphi(x)+\varphi(x-\Delta x)}{(\Delta x)^2}+O[(\Delta x)^2] \tag{3.11}$$

If the remainder term $O[(\Delta x)^2]$ was omitted, we can get:

$$\frac{\partial^2 \varphi}{\partial x^2}\approx\frac{\varphi(x+\Delta x)-2\varphi(x)+\varphi(x-\Delta x)}{(\Delta x)^2} \tag{3.12}$$

Equation (3.12) is the central-difference formula of second-order derivative, the truncation error is also second-order.

The above-mentioned difference formula of the derivative is for the independent variable x. Similarly, if the head $\varphi(x)$ is considered as a function of the spatial variable y or the time variable t, we can also get corresponding difference formulae.

3.2.2 Convergence and Stability

Convergence and stability are the terms that describe the accuracy of a finite difference solution of the corresponding problems.

1. *Convergence*

In terms of a partial differential equation with independent variables x, y, and t. Let U represent the exact solution to the partial differential equation. Now let u be the exact solution to the finite difference equations that are used to approximate the partial differential equation. The finite difference equations are said to be convergent if $u \to U$ as Δx, Δy and $\Delta t \to 0$. The difference between U and u is referred to as the truncation or discretization error. Discretization error is incurred by truncating terms of equations (3.3) and (3.8) in developing the finite difference analogues to the partial derivatives.

2. *Stability*

When the finite difference equations are solved, roundoff errors are generated because the calculation cannot be carried out to an infinite number of significant figures. The calculated numerical solution can be called $\boldsymbol{S}$. If $|\varepsilon_{i,j}^{k}|$ represents the absolute value of the computational error at the (i, j) node for k^{th} time interval and all are less than ε, then the finite difference equations are said to be stable if $\boldsymbol{S} \to \boldsymbol{u}$ as $\varepsilon \to 0$. ε would tend toward zero as more significant figures were used in the computations.

§ 3.3 Steady Flow in Confined Aquifers

3.3.1 1-D Steady Flow

3.3.1.1 In Homogeneous Confined Aquifer

Question: Assuming a confined aquifer bounded with two parallel rivers, it is homogeneous, isotropic, equal-thickness and has no vertical recharge. The distance between the two rivers is L, and the river levels are H_0 and H_L respectively (Figure 3.2). Please solve the variation rule of the heads in this aquifer.

1. *Conceptual hydrogeological model*

(1) Confined aquifer.

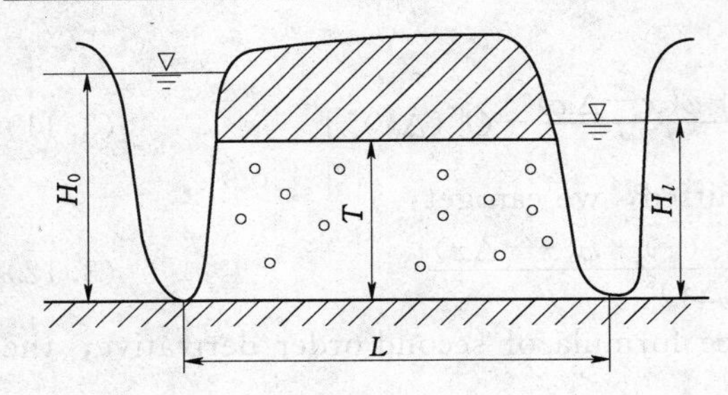

Figure 3. 2 1-D steady flow in confined aquifer between rivers

(2) Homogeneous porous medium (T is constant).

(3) Bounded with two rivers.

(4) One - Dimension (1 - D) steady flow.

2. *Mathematical model*

$$\begin{cases} T\dfrac{\partial^2 \varphi}{\partial x^2}=0 & 0<x<l \\ \varphi|_{x=0}=H_0 \\ \varphi|_{x=l}=H_l \end{cases} \tag{3. 13}$$

3. *Analytical solution*

The analytical solution as follows:

Groundwater head:

$$\varphi=H_0-\frac{H_0-H_l}{l}x \tag{3. 14}$$

Total discharge:

$$Q=T\frac{H_0-H_l}{l} \tag{3. 15}$$

Specific discharge:

$$q=K\frac{H_0-H_l}{l} \tag{3. 16}$$

Average velocity:

$$v=\frac{q}{n_e} \tag{3. 17}$$

Travel time:

$$t=\int_0^l\frac{\mathrm{d}x}{v}=\int_0^l\frac{\mathrm{d}x}{q/n_e}=\frac{n_e}{K}\frac{l^2}{H_0-H_l} \tag{3. 18}$$

$$t=\frac{l}{v}=\frac{l}{q/n_e}=\frac{n_e l}{q} \tag{3. 19}$$

There is an example:

H_0 (m)	H_1 (m)	M (m)	K (m/d)	T (m²/d)	l (m)	n_e	Q (m³/d)	q (m/d)	T (d)
20	14	10	10	100	1000	0. 2	0. 6	0. 06	3333
		10	100	1000			6	0. 6	333
		100	10	1000			6	0. 06	3333

4. *Finite difference equations*

Figure 3. 3 shows the grid dicretization.

Discretization:

$$\Delta x=\frac{l}{n};x_i=i\Delta x;\varphi_i=\varphi(i\Delta x)$$

Water balance:

$$\frac{\varphi_{i-1}-\varphi_i}{\Delta x}T+\frac{\varphi_{i+1}-\varphi_i}{\Delta x}T=0 \qquad (3.20)$$

Finite difference equation:

$$-\varphi_{i-1}+2\varphi_i-\varphi_{i+1}=0 \qquad (3.21)$$

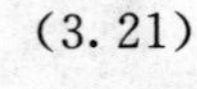

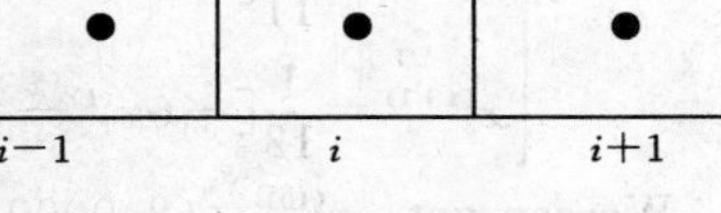

Figure 3.3 Grid discretization

5. *Iteration solutions*

(1) Jacobi iteration.

$$\varphi_i^{(m+1)}=\frac{\varphi_{i-1}^{(m)}+\varphi_{i+1}^{(m)}}{2} \qquad (3.22)$$

Where: m is iteration index.

Solution procedure:

Step 1 Assuming any initial values$\varphi_i^{(0)}$ $(i=1,2,\cdots,n)$.

Step 2 Using equation (3.22) to compute $\varphi_i^{(m)}$ $(i=1,2,\cdots,n;m=1,2,3,\cdots)$.

Notice! Using values from previous iteration cycle at all nodes.

Step 3 Repeat step 2 until

$$\text{maximum } |\phi_i^{m+1}-\phi_i^{(m)}|<\varepsilon \quad i=1,2,\cdots,n \quad \text{or} \quad m=M_{\max} \qquad (3.23)$$

Where: ε is the convergence criterion; $M_{\max}$ is the maximum number of iterations.

In order to further explain the principle and method of Jacobi iteration and Guess—Siedel iteration, we consider the linear simultaneous equation, such as $A\cdot x=b$.

$$\begin{cases}8x_1-3x_2+2x_3=20\\4x_1+11x_2+x_3=33\\6x_1+3x_2+12x_3=36\end{cases}$$

Here

$$A=\begin{pmatrix}8&-3&2\\4&11&1\\6&3&12\end{pmatrix}\quad x=\begin{pmatrix}x_1\\x_2\\x_3\end{pmatrix}\quad b=\begin{pmatrix}20\\33\\36\end{pmatrix}$$

Therefore, the exact solution is $x=(3,\ 2,\ 1)$.

Then it is adapted as follows:

$$\begin{cases}x_1=\frac{1}{8}(3x_2-2x_3+20)\\x_2=\frac{1}{11}(-4x_1-x_3+33)\\x_3=\frac{1}{12}(-6x_1-3x_2+36)\end{cases}$$

Or be adapted as $x=B_0x+f$.

Here

$$B_0=\begin{pmatrix}0&\frac{3}{8}&-\frac{2}{8}\\-\frac{4}{11}&0&-\frac{1}{11}\\-\frac{1}{2}&-\frac{1}{4}&0\end{pmatrix}\qquad f=\begin{pmatrix}\frac{5}{2}\\3\\3\end{pmatrix}$$

Taking initial value $x^{(0)}=(0,\ 0,\ 0)^{\mathrm{T}}$, it will be brought into the equation and obtained the new solution $x^{(k)}=[x_1^{(k)}+x_2^{(k)}+x_3^{(k)}]^{\mathrm{T}}$.

$$\begin{cases} x_1^{(k+1)} = \frac{1}{8}[3x_2^{(k)} - 2x_3^{(k)} + 20] \\ x_2^{(k+1)} = \frac{1}{11}[-4x_1^{(k)} + x_3^{(k)} + 33] \\ x_3^{(k+1)} = \frac{1}{12}[-6x_1^{(k)} - 3x_2^{(k)} + 36] \end{cases} \quad \text{or simplified as } x^{(k+1)} = B_0 x^{(k)} + f$$

We can get: $x^{(10)} = (3.000032, 1.999838, 0.9998813)^T$.

Therefore, the formula of Jacobi iteration is simple; it only requires to calculate the multiplication of matrix and vector once at every iterations. During the calculation, the original matrix A keeps constant, instead of using the newly computed values of the variable. So, its convergence rate is slower and the calculation is relatively longer.

(2) Guess—Siedel iteration.

$$\varphi_i^{(m+1)} = \frac{\varphi_{i-1}^{(m+1)} + \varphi_{i+1}^{(m)}}{2} \tag{3.24}$$

Solution procedure:

Step 1 Assuming any initial values $\varphi_i^{(0)}$ $(i=1,2,3,\cdots,n)$.

Step 2 Applying equation (3.24) to nodes from left to right to compute $\phi_i^{(m)}$ using the newly computed $\phi_{i-1}^{(m+1)}$ $(i=1, 2, 3, \cdots, n; m=1, 2, 3, \cdots)$.

Step 3 Repeat step 2 for $m=1, 2, 3, 4, \cdots$ until the convergence criterion is met or maximum number of iterations is reached.

Guess—Siedel iteration method is an improvement of Jacobi iteration, its character is that each element is calculated by using the newly computed values. Therefore, its convergence rate is fast. Considering the linear simultaneous equation mentioned above $A \cdot x = b$, we can get Guess—Siedel iteration:

$$\begin{cases} x_1^{(k+1)} = \frac{1}{8}[3x_2^{(k)} - 2x_3^{(k)} + 20] \\ x_2^{(k+1)} = \frac{1}{11}[-4x_1^{(k+1)} + x_3^{(k)} + 33] \\ x_3^{(k+1)} = \frac{1}{12}[-6x_1^{(k+1)} - 3x_2^{(k+1)} + 36] \end{cases}$$

We can get: $x^{(7)} = (3.000002, 1.999987, 0.9999932)^T$.

6. *Direct solution*

Applying the finite difference equation (3.21) to all nodes will result in:

$$\begin{array}{lll} i=1 & 2\varphi_1 - \varphi_2 & = H_0 \\ i=2 & -\varphi_1 + 2\varphi_2 - \varphi_3 & = 0 \\ i=3 & \quad -\varphi_2 + 2\varphi_3 - \varphi_4 & = 0 \\ \vdots & & \\ i=n-2 & \qquad -\varphi_{n-3} + 2\varphi_{n-2} - \varphi_{n-1} & = 0 \\ i=n-1 & \qquad\quad -\varphi_{n-2} + 2\varphi_{n-1} & = H_l \end{array} \tag{3.25}$$

There are $n-1$ equations for $n-1$ unknown φ_i $(i=1, 2, \cdots, n-1)$.

Equation (3. 25) can be rewritten as:

$$\begin{bmatrix} 2 & -1 & & & & & \varphi_1 & & & & \\ & -1 & 2 & -1 & & & & \varphi_2 & & & \\ & & -1 & 2 & -1 & & & & \varphi_3 & & \\ & \vdots & & & & & & & & & \\ & & & \cdots & & & & & & & \\ & & & \vdots & & & & & & & \\ & & & & -1 & 2 & -1 & & & \varphi_{n-2} & \\ & & & & & -1 & 2 & & & & \varphi_{n-1} \end{bmatrix} = \begin{bmatrix} H_0 \\ 0 \\ 0 \\ \vdots \\ 0 \\ H_l \end{bmatrix} \tag{3. 26}$$

Equation (3. 26) is a well known tri-diagonal system.

7. *Solution of tri-diagonal system equation (Thomas method)*

Equation (3. 26) can be generalized in the form:

$$\begin{vmatrix} b_1 & c_1 & & & & \\ a_2 & b_2 & c_2 & & & \\ & a_3 & b_3 & c_3 & & \\ & & \vdots & \vdots & \vdots & \\ & & & \vdots & \vdots & \vdots \\ & & & a_{n-1} & b_{n-1} & c_{n-1} \\ & & & & a_n & b_n \quad c_n \end{vmatrix} \times \begin{vmatrix} x_1 \\ x_2 \\ x_3 \\ \vdots \\ x_{n-1} \\ x_n \end{vmatrix} = \begin{vmatrix} f_1 \\ f_2 \\ f_3 \\ \vdots \\ f_{n-1} \\ f_n \end{vmatrix} \tag{3. 27}$$

Or in matrix form:

$$AX=f \tag{3. 28}$$

In the equation (3. 28), A has a unique decomposition LU when met the following conditions.

When $|i-j|>1, a_{ij}=0$ and

(1) $|b_1|>|c_1|>0$.

(2) $|b_i|\geqslant|a_i|+|c_i| \quad a_i, c_i\neq 0(i=2,3,\cdots,n-1)$.

(3) $|b_n|>|a_n|>0$.

A can be factorized into the product of one lower triangular matrix (L) and one upper triangular matrix (U):

$$A=LU=\begin{vmatrix} \alpha_1 & & & & \\ a_2 & \alpha_2 & & & \\ & a_3 & \alpha_3 & & \\ & \vdots & \vdots & & \\ & & \vdots & \vdots & \\ & & & \vdots & \vdots \\ & & & a_n & \alpha_n \end{vmatrix} \begin{vmatrix} 1 & \beta_1 & & & \\ & 1 & \beta_2 & & \\ & & 1 & \beta_3 & \\ & & & \vdots & \vdots \\ & & & & \vdots \\ & & & \vdots & \beta_{n-1} \\ & & & & 1 \end{vmatrix} \tag{3. 29}$$

Since elements in LU equal to elements in A, we get:

$$\begin{cases} \alpha_1=b_1 & \\ \alpha_i\beta_i=c_i & i=1,2,\cdots,n-1 \\ a_i\beta_{i-1}+\alpha_i=b_i & i=2,3,\cdots,n \end{cases} \tag{3. 30}$$

Solve equation (3.30) for β_i and α_i:

$$\begin{cases} \alpha_1 = b_1 \\ \beta_i = \dfrac{c_i}{\alpha_i} & i=1,2,\cdots,n-1 \\ \alpha_i = b_i - a_i\beta_{i-1} & i=2,3,\cdots,n \end{cases} \tag{3.31}$$

The matrices L and U are defined.

Equation (3.28) becomes:

$$\boldsymbol{LU}x = f \tag{3.32}$$

Let $Ux=y$, then $Ly=f$.

Solution procedure of Thomas method:

Step 1 Work out n and Δx.

Step 2 Applying equation (3.31) to compute β_i and α_i.

Step 3 $\boldsymbol{Ly=f}$ is solved by foreward substitution:

$$\begin{cases} y_1 = \dfrac{f_1}{\alpha_1} \\ y_i = \dfrac{f_i - a_i y_{i-1}}{\alpha_i} & i=2,3,\cdots,n \end{cases} \tag{3.33}$$

y_i is solved by equation (3.33) ($i=2, \cdots, n$), step 2 and 3 is called "chasing".

Step 4 $Ux=y$ is solved by backward substitution:

$$\begin{cases} x_n = y_n \\ x_i = y_i - \beta_i x_{i+1} & i=n-1,n-2,\cdots,1 \end{cases} \tag{3.34}$$

x_i is solved by equation (3.34) ($i=n-1, \cdots, 1$), this process is called "rushing", x_i is the heads for the problem.

8. *Table for calculation*

Node number	A			f	LU		y_i	x_i	φ_i
i	a_i	b_i	c_i	f_i	α_i	β_i			
1	0	2	−1	20	2	−0.5	10	19	19
2	−1	2	−1	0	1.5	−1/1.5	20/3	18	18
3	−1	2	−1	0	2/1.5	−1.5/2	5	17	17
4	−1	2	−1	0	2.5/2	−2/2.5	4	16	16
5	−1	2	0	14	3/2.5	−2.5/3	15	15	15

Exercise

A confined aquifer bounded with two parallel rivers. The width and the thickness of the aquifer is 2000m and 20m respectively. The level of left river is 30m, the right is 18m. The transmissivity is 100m^2/d. Try to calculate the heads of groundwater using Thomas method ($n=5$).

3.3.1.2 In Non-homogeneous Confined Aquifer

Question: Assuming a confined aquifer bounded with two parallel rivers, it is non-homogeneous and has no vertical recharge. The distance between the two rivers is L, and the river levels are H_0 and H_l respectively (Figure 3.4). Please solve the variation of the heads in this aquifer.

1. *Conceptual hydrogeological model*

(1) Confined aquifer.

(2) Inhomogeneous porous medium (T is a function of x).

(3) Bounded with two rivers.

(4) 1-D steady flow.

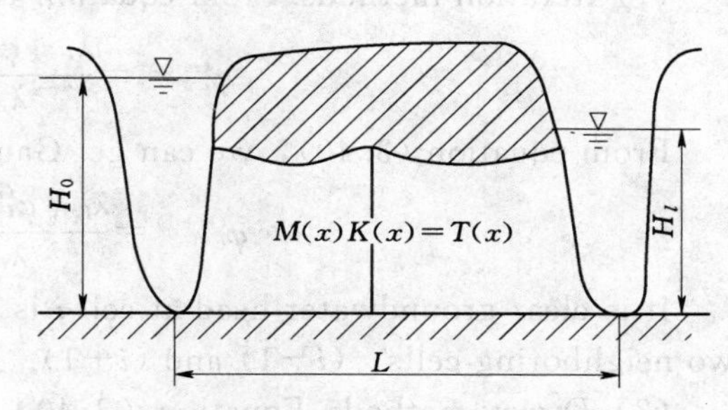

Figure 3.4 1-D steady flow in non-homogeneous confined aquifer between rivers

2. *Mathematical model*

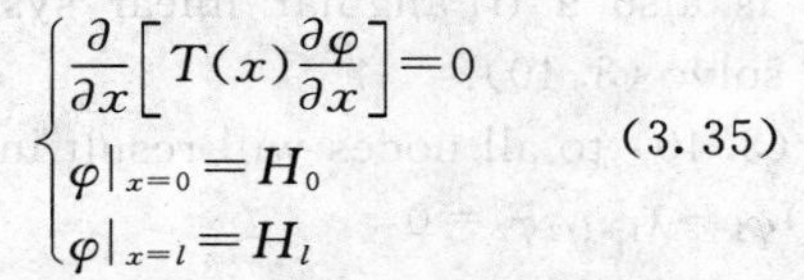

$$\begin{cases} \dfrac{\partial}{\partial x}\left[T(x)\dfrac{\partial \varphi}{\partial x}\right]=0 \\ \varphi|_{x=0}=H_0 \\ \varphi|_{x=l}=H_l \end{cases} \tag{3.35}$$

There is no analytical solution exists for equation (3.35).

3. *Finite difference equations*

(1) Discretization.

$i-1$	i	$i+1$
$T_{i-\frac{1}{2}}$, $M_{i-\frac{1}{2}}$, $K_{i-\frac{1}{2}}$ (between $i-1$ and i)		$T_{i+\frac{1}{2}}$, $M_{i+\frac{1}{2}}$, $K_{i+\frac{1}{2}}$ (between i and $i+1$)
φ_{i-1}	φ_i	φ_{i+1}
Δx_{i-1}	Δx_i	Δx_{i+1}
t_{i-1}	t_i	t_{i+1}

(2) Water balance equation.

$$\frac{\varphi_{i-1}-\varphi_i}{\Delta x_{i-\frac{1}{2}}}T_{i-\frac{1}{2}}+\frac{\varphi_{i+1}-\varphi_i}{\Delta x_{i+\frac{1}{2}}}T_{i+\frac{1}{2}}=0 \tag{3.36}$$

Let

$$\lambda_{i-\frac{1}{2}}=\frac{T_{i-\frac{1}{2}}}{\Delta x_{i-\frac{1}{2}}} \quad \lambda_{i+\frac{1}{2}}=\frac{T_{i+\frac{1}{2}}}{\Delta x_{i+\frac{1}{2}}} \tag{3.37}$$

Where

$$T_{i-\frac{1}{2}}=\frac{2T_{i-1}T_i}{T_{i-1}+T_i} \quad T_{i+\frac{1}{2}}=\frac{2T_iT_{i+1}}{T_i+T_{i+1}} \tag{3.38}$$

$$\Delta x_{i-\frac{1}{2}}=\frac{\Delta x_{i-1}+\Delta x_i}{2} \quad \Delta x_{i+\frac{1}{2}}=\frac{\Delta x_i+\Delta x_{i+1}}{2} \tag{3.39}$$

(3) The finite difference equation will be

$$-\lambda_{i-\frac{1}{2}}\varphi_{i-1}+(\lambda_{i-\frac{1}{2}}+\lambda_{i+\frac{1}{2}})\varphi_i-\lambda_{i+\frac{1}{2}}\varphi_{i+1}=0 \tag{3.40}$$

4. *Solutions*

Both Jacobi and Gauss—siedel method can be applied to equation (3.40).

(1) Iteration methods. From equation (3. 40), we can get Jacobi iteration:

$$\varphi_i^{(m+1)} = \frac{\lambda_{i-\frac{1}{2}}\varphi_{i-1}^{(m)} + \lambda_{i+\frac{1}{2}}\varphi_{i+1}^{(m)}}{\lambda_{i-\frac{1}{2}} + \lambda_{i+\frac{1}{2}}} \tag{3. 41}$$

From equation (3. 40), we can get Gauss—siedel iteration:

$$\varphi_i^{(m+1)} = \frac{\lambda_{i-\frac{1}{2}}\varphi_{i-1}^{(m+1)} + \lambda_{i+\frac{1}{2}}\varphi_{i+1}^{(m)}}{\lambda_{i-\frac{1}{2}} + \lambda_{i+\frac{1}{2}}} \tag{3. 42}$$

It is clear groundwater head at cell i is a weighted average of groundwater heads in two neighboring cells: $(i-1)$ and $(i+1)$.

(2) Direct methods. Equation (3. 40) is also a tri-angular linear systems equation. The Thomas method can be applied to solve (3. 40).

Applying the finite difference equation (3. 40) to all nodes will result in:

$$\begin{cases} -\lambda_{1/2}H_0 + (\lambda_{1/2} + \lambda_{1+1/2})\varphi_1 - \lambda_{1+1/2}\varphi_2 = 0 \\ -\lambda_{2-1/2}\varphi_1 + (\lambda_{2-1/2} + \lambda_{2+1/2})\varphi_2 - \lambda_{2+1/2}\varphi_3 = 0 \\ \vdots \\ -\lambda_{n-5/2}\varphi_{n-3} + (\lambda_{n-5/2} + \lambda_{n-3/2})\varphi_{n-2} - \lambda_{n-3/2}\varphi_{n-1} = 0 \\ -\lambda_{n-3/2}\varphi_{n-2} + (\lambda_{n-3/2} + \lambda_{n+1/2})\varphi_{n-1} - \lambda_{n-1/2}H_1 = 0 \end{cases}$$

Exercises

A confined aquifer bounded with two parallel rivers, the width and the thickness of the aquifer is 1000m and the thickness is $M(x) = 20 - 0.01x$. The level of left river is 30m, the right is 18m. The hydraulic conductivity is 10m/d. Try to calculate the heads of groundwater using finite difference method (three kinds of solutions) for 1 - D steady flow.

3. 3. 2 2-D Steady Flow

In order to explain the method of finite difference for 2-D steady groundwater flow, we assume that the confined aquifer is homogeneous, isotropic, equal-thickness and has no vertical recharge. The vadose zone is rectangular and its upper boundary and the lower left corner boundary can be treated as first type boundaries, rest of the boundaries can be treated as second type boundaries. The upper boundary is $H_0 = 100$m and the lower left corner is $H_0 = 50$m (Figure 3. 5). Please solve the variation of the heads in this aquifer.

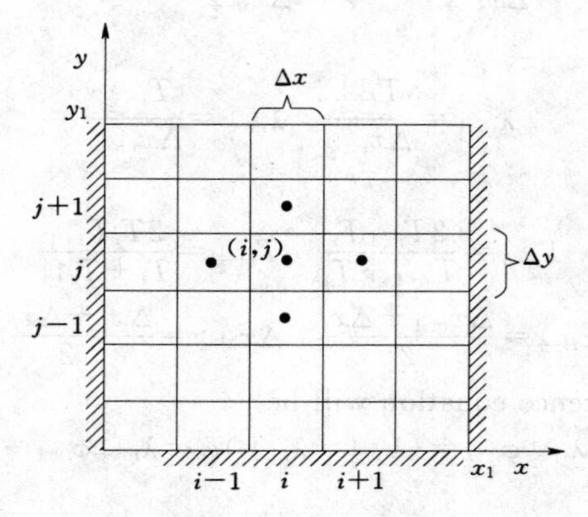

Figure 3. 5 2-D steady flow

1. *Conceptual hydrogeological model*

(1) Rectangular region.

(2) Homogeneous aquifer (T is constant).

(3) Upper boundary $H_0=100\text{m}$, lower left corner $H_0=50\text{m}$; left and right boundaries are impermeable.

(4) Two - Dimension (2 - D) steady flow.

2. *Mathematical model*

$$\begin{cases} \dfrac{\partial^2\varphi}{\partial x^2}+\dfrac{\partial^2\varphi}{\partial y^2}=0 & 0<x<x_l;0<y<y_l \\ \varphi|_{y=0}=50 & 0<x<x_0 \\ \varphi|_{y=y_l}=100 & x_0<x<x_l \\ \left.\dfrac{\partial\varphi}{\partial x}\right|_{x=0}=\left.\dfrac{\partial\varphi}{\partial x}\right|_{x=x_l}=0 & 0<y<y_l \\ \left.\dfrac{\partial\varphi}{\partial y}\right|_{y=0}=0 & 0<x<x_l \end{cases} \tag{3.43}$$

3. *Finite-difference equations*

Y direction:

$$\begin{cases} (i,j-1)\Rightarrow(i,j) & T\dfrac{\varphi_{i,j-1}-\varphi_{i,j}}{\Delta y}\Delta x \\ (i,j+1)\Rightarrow(i,j) & T\dfrac{\varphi_{i,j+1}-\varphi_{i,j}}{\Delta y}\Delta x \end{cases}$$

X direction:

$$\begin{cases} (i+1,j)\Rightarrow(i,j) & T\dfrac{\varphi_{i+1,j}-\varphi_{i,j}}{\Delta x}\Delta y \\ (i-1,j)\Rightarrow(i,j) & T\dfrac{\varphi_{i-1,j}-\varphi_{i,j}}{\Delta x}\Delta y \end{cases}$$

Let $\Delta x=\Delta y$; Water balance equation will be:

$$\varphi_{i,j-1}+\varphi_{i,j+1}-4\varphi_{i,j}+\varphi_{i-1,j}+\varphi_{i+1,j}=0 \tag{3.44}$$

Or

$$\varphi_{i,j}=\frac{1}{4}(\varphi_{i-1,j}+\varphi_{i+1,j}+\varphi_{i,j-1}+\varphi_{i,j+1}) \tag{3.45}$$

The value of head $\varphi_{i,j}$ at node (i, j) equals to the average value of heads at the four neighboring nodes.

Note:

(1) There is no need for equation on the first type of boundary, since $\varphi_{i,j}$ is known.

(2) On the impermeable boundary (y), assuming a imaginary node outside of the boundary with a distance of $\Delta x/2$, boundary condition: $\dfrac{\partial\varphi}{\partial x}=0$.

When give: $T\dfrac{\varphi_{i+1,j}-\varphi_{i,j}}{\Delta x}=0$, so $\varphi_{i+1,j}=\varphi_{i,j}$.

From equation (3.45), the water balance equation for (i, j) becomes:

$$\varphi_{i,j}=\frac{1}{3}(\varphi_{i-1,j}+\varphi_{i,j-1}+\varphi_{i,j+1}) \tag{3.46}$$

It is the average of heads in the three neighboring nodes.

(3) If the nodes are on the horizontal impermeable boundaries, assuming an imaginary node outside of impermeable boundary with a distance of Δy and applying boundary condition gives:

$$T\frac{\varphi_{i,j-1}-\varphi_{i,j+1}}{2\Delta y}=0 \quad \text{so} \quad \varphi_{i,j-1}=\varphi_{i,j+1}$$

From equation (3.45), the water balance equation for (i, j) becomes:

$$\varphi_{i,j}=\frac{1}{4}(\varphi_{i-1,j}+\varphi_{i+1,j}+2\varphi_{i,j+1}) \tag{3.47}$$

4. *Iteration solutions*

(1) Jacob iteration.

$$\varphi_{i,j}^{(m+1)}=\frac{\varphi_{i-1,j}^{(m)}+\varphi_{i+1,j}^{(m)}+\varphi_{i,j-1}^{(m)}+\varphi_{i,j+1}^{(m)}}{4} \tag{3.48}$$

Where: m is iteration index.

Solution procedures are:

Step 1 Assuming any initial value $\varphi_{i,j}^{(0)}$ $(i=1, 2, \cdots, N; j=1, 2, \cdots, M)$.

Step 2 Using equation (3.48) to compute $\varphi_{i,j}^{(m)}$ for $m=1, 2, 3, \cdots$. At all nodes using values from previous iteration cycle.

Step 3 Repeat step 2 until

$$\mathrm{Max}|\varphi_{i,j}^{(m+1)}-\varphi_{i,j}^{(m)}|<\varepsilon \quad (i=1,2,\cdots,N;\ j=1,2,\cdots,M) \tag{3.49}$$

The drawback of Jacobi iteration is the slow convergence rate.

(2) Gauss—Siedel iteration. Whatever any node, using the newly computed value (mth) at its neighboring nodes as a trial to compute the improved value $(m+1)$ th with Jacobi iteration. The advantage of Gauss—Siedel iteration is the utilization of the newly improved values at neighboring nodes in time. The iteration equation shown as following:

$$\varphi_{i,j}^{(m+1)}=\frac{\varphi_{i-1,j}^{(m+1)}+\varphi_{i,j-1}^{(m+1)}+\varphi_{i+1,j}^{(m)}+\varphi_{i,j+1}^{(m)}}{4} \tag{3.50}$$

Solution procedures are:

Step 1 Give any initial values $\varphi_{i,j}^{(0)}$.

Step 2 Apply equation (3.50) to nodes from left to right on a row by row sequence, staring from node (1, 1), using newly computed $\varphi_{i-1,j}^{(m+1)}$, $\varphi_{i,j-1}^{(m+1)}$ to compute $\varphi_{i,j}^{(m+1)}$.

Step 3 Repeat step 2 for $m=1, 2, 3, \cdots$until the convergence criterion is met.

Gauss—Siedel iteration is one of the common methods, has a higher convergence rate than Jacobi iteration.

(3) Successive over-relaxation (SOR) method. The SOR method is a kind of linear acceleration iteration methods. This method is linear combination the last improved values $h_{i,j}^{(m)}$ and the Gauss—Seidel iteration values $\tilde{h}_{i,j}^{(m+1)}$ appropriately to form an iterative formula with higher convergence rate.

The SOR method uses an intermediate value computed by equation (3.50) denoted as

$$\varphi_{i,j}^{(m+1)}=\frac{\varphi_{i-1,j}^{(m+1)}+\varphi_{i,j-1}^{(m+1)}+\varphi_{i+1,j}^{(m)}+\varphi_{i,j+1}^{(m)}}{4} \tag{3.51}$$

Then $\varphi_{i,j}^{(m+1)}$ is formed by extrapolation

$$\varphi_{i,j}^{(m+1)}=\varphi_{i,j}^{(m)}+\omega(\varphi_{i,j}^{(m+1)}-\varphi_{i,j}^{(m)}) \tag{3.52}$$

ω is called over-relaxation factor or acceleration parameter ranging from $1<\omega<2$. If we select a proper over-relaxation factor, the convergence rate can be speeded up markedly. However, the values of ω can only be selected empirically at present.

Equation (3.52) shows that a better estimate $\varphi_{i,j}^{(m+1)}$ is obtained with $\varphi_{i,j}^{(m)}$ by considering the error $(\varphi_{i,j}^{(m+1)}-\varphi_{i,j}^{(m)})$. The error is relaxed with ω.

Since the sequence of computation using equation (3.51) and equation (3.52) is node by node and row by row, starting from node (1, 1). Therefore, the speed of convergence is in order of SOR>Gauss—Siedel>Jacob.

SOR can be also applied to transient flow problems.

§ 3.4 Transient Flow in Confined Aquifers

3.4.1 1-D Transient Flow

To illustrate the specific steps for solving the problems of 1-D transient groundwater flow in confined aquifers using the finite difference method, please consider the following example.

As shown in Figure 3.6, the confined aquifer is homogeneous, isotropic, equal-thickness and has no vertical recharge. The initial groundwater head (when $t=0$) is H_0 in the aquifer. When $t>0$, left river level is increased by ΔH. please solve the variation of the heads in this aquifer.

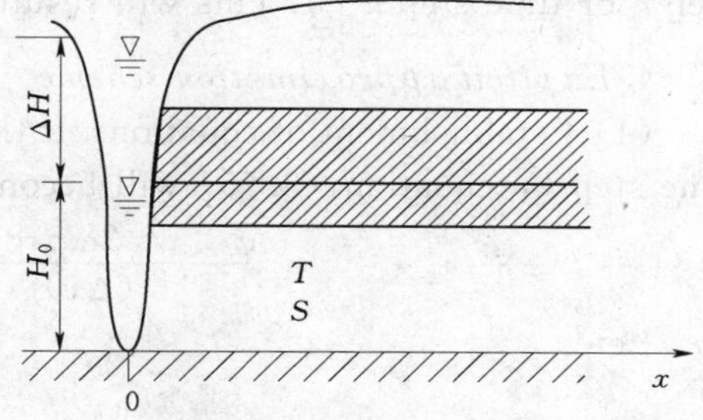

Figure 3.6 1-D transient flow in confined aquifer

1. *Conceptual hydrogeological model*

(1) Homogeneous confined aquifer: transmissivity ($T=KM$) and storage coefficient (S) are constant.

(2) Aquifer is bounded by a river on the left and extends to infinity on the right.

(3) Initial groundwater head (when $t=0$) is horizontal (H_0).

(4) Left river stage is increased by ΔH when $t>0$.

2. *Mathematical model*

$$\begin{cases}\dfrac{\partial^2\varphi}{\partial x^2}=\dfrac{S}{T}\dfrac{\partial\varphi}{\partial t}\\ \varphi|_{x=0}=H_0+\Delta H\delta(t) \quad t>0\\ \varphi|_{t=0}=H_0 \quad X>0\\ \varphi|_{x=+\infty}=H_0\end{cases} \tag{3.53}$$

3. *Finite difference equations*

(1) Discretization. As shown in Figure 3.7.

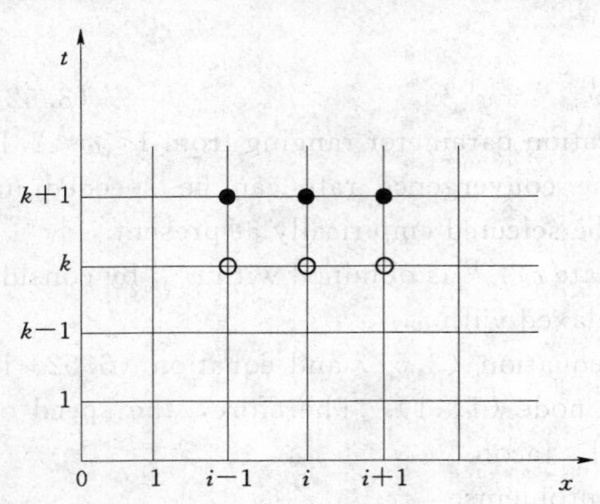

Figure 3.7 Zone discretization

1-D space: Δx $x_i = i\Delta x$, $i = 0, 1, \cdots, n$.

Time: Δt $t_k = k\Delta t$, $k = 0, 1, \cdots, m$

(2) Finite Difference Approximation. From the finite-difference approximation of second-order derivative we can get

$$\begin{cases} T\dfrac{\partial^2 \varphi}{\partial x^2} \approx T\dfrac{\varphi_{i+1} - 2\varphi_i + \varphi_{i-1}}{(\Delta x)^2} \\ S\dfrac{\partial \varphi}{\partial t} = S\dfrac{\varphi_{i,k+1} - \varphi_{i,k}}{\Delta t} \end{cases} \tag{3.54}$$

According to the equation (3.53), equation (3.54) can be re-arranged as

$$\frac{\varphi_{i-1} - 2\varphi + \varphi_{i+1}}{(\Delta x)^2} = \frac{S}{T}\frac{\varphi_{i,k+1} - \varphi_{i,k}}{\Delta t} \tag{3.55}$$

Comparing equation (3.55) with the partial differential equation in equation (3.54) can be found and the left side in equation (3.55) is the approximation of spatial differential and the right side is the approximation of time differential. However, the choice has to be made on whether the spatial difference in equation (3.55) should be taken at time step k or time step $k+1$. This will result in different finite difference scheme.

4. *Explicit approximation scheme*

(1) Establishment of equations. If the difference in equation (3.55) will be taken at time step k, equation (3.55) will become

$$\frac{\varphi_{i-1,k} - 2\varphi_{i,k} + \varphi_{i+1,k}}{(\Delta x)^2} = \frac{S}{T}\frac{\varphi_{i,k+1} - \varphi_{i,k}}{\Delta t} \tag{3.56}$$

Let

$$\gamma = \frac{T\Delta t}{S(\Delta x)^2} \tag{3.57}$$

The solution for $\varphi_{i,k+1}$ will be:

$$\varphi_{i,k+1} = \gamma\varphi_{i-1,k} + (1-2\gamma)\varphi_{i,k} + \gamma\varphi_{i+1,k} \tag{3.58}$$

It is clear that for $k = 0, 1, 2, \cdots, m$, the unknown head $\varphi_{i,k+1}$ is represented explicitly by three known heads. Therefore, equation (3.57) is called **the explicit finite difference scheme.**

(2) Solution procedures.

When $k=0$

$$\varphi_{i,1} = \gamma\varphi_{i-1,0} + (1-2\gamma)\varphi_{i,0} + \gamma\varphi_{i+1,0}$$

Where: $\varphi_{i,0} = H_0 (i = 1, 2, \cdots, n-1)$ are initial conditions.

When $k=1$

$$\varphi_{i,2} = \gamma\varphi_{i-1,1} + (1-2\gamma)\varphi_{i,1} + \gamma\varphi_{i+1,1}$$

The boundary conditions: $\varphi_{0,k} = H_0 + \Delta H$, $\varphi_{n,k} = H_0 (k = 0, 1, 2, \cdots, m-1)$ will be used. In this way, the solution is "marched" forward in time.

Example: About upper problem, we assume: $T=1000\text{m}^2/\text{d}$, $S=0.01$, $H_0=14\text{m}$, $\Delta H=6\text{m}$, $L=1000\text{m}$. Choose $\Delta x=100\text{m}$ and $\Delta t=0.01\text{d}$.

Thus: $\gamma=\dfrac{T\Delta t}{S(\Delta x)^2}=\dfrac{1000\times 0.01}{0.01\times 10000}=0.1$.

And equation (3. 58) will become

$$\varphi_{i,k+1}=0.1\varphi_{i-1,k}+0.8\varphi_{i,k}+0.1\varphi_{i+1,k} \tag{3.59}$$

Use equation (3. 59) to compute the groundwater heads up to $k=16$.

The groundwater heads at $i=1, 2, \cdots, 9$ will gradually increase.

If we choose $\Delta t=0.1\text{d}$, thus $\gamma=1$, and equation (3. 59) will become

$$\varphi_{i,k+1}=\varphi_{i-1,k}-\varphi_{i,k}+\varphi_{i+1,k} \tag{3.60}$$

We can use equation (3. 60) to compute groundwater heads with the same table. However, you will find that the calculation is not stable.

Table of calculation

k \ i	0	1	2	3	4	5	6	7	8	9	10
	0	100	200	300	400	500	600	700	800	900	1000
0	14										14
1	20										14
2	20										14
3	20										14
⋮	20										14
16	20										14

(3) Stability and convergence of Explicit Approximation Scheme. Explicit Approximation Scheme will be stable and convergent, if and only if $0<\gamma\leqslant 1/2$.

(4) Disadvantages. Good approximation: Δx must be small, if (Δx) is small, Δt will be required very small in order to satisfy $\gamma\leqslant 1/2$. Therefore, it may need extremely large number of time steps to simulate a short time period.

5. *Implicit approximation scheme*

(1) Establishment of equations. If the spatial difference in equation (3. 55) is taken at time step $k+1$, it will become

$$\frac{\varphi_{i-1,k+1}-2\varphi_{i,k+1}+\varphi_{i+1,k+1}}{(\Delta x)^2}=\frac{S}{T}\frac{\varphi_{i,k+1}-\varphi_{i,k}}{\Delta t} \tag{3.61}$$

Use the same $\gamma=\dfrac{T\Delta t}{S(\Delta x)^2}$.

The solution for $\varphi_{i,k+1}$ will be:

$$\gamma\varphi_{i-1,k+1}-(1+2\gamma)\varphi_{i,k+1}+\gamma\varphi_{i-1,k+1}=-\varphi_{i,k} \tag{3.62}$$

Equation (3. 62) is called the implicit finite difference scheme since there are three unknown heads.

(2) Solution procedures.

Step 1 Start from $k=0$, equation (3. 62) will become

$$-\gamma\varphi_{i-1,1}+(1+2\gamma)\varphi_{i,1}-\gamma\varphi_{i-1,1}=-\varphi_{i,0} \tag{3.63}$$

Using boundary conditions$\varphi_{0,k}=H_0+\Delta H$ ($k=0$, 1, 2, …, $m-1$) and initial conditions $\varphi_{i,0}=H_0$ ($i=1$, 2, …, $n-1$), equation (3.63) is a tri-diagonal systems equation and can be solved with Thomas or iteration method for $\varphi_{i,1}$ ($i=1$, 2, …, $n-1$).

Step 2 With known $\varphi_{i,1}$, solve the equation (3.64) for $\varphi_{i,2}$.

$$-\gamma\varphi_{i-1,2}+(1+2\gamma)\varphi_{i,2}-\gamma\varphi_{i-1,2}=-\varphi_{i,1} \tag{3.64}$$

The procedure can be repeated up to time step $m-1$.

The solution is marched forward in time by solving the system of equation at each time-step.

(3) Stability and convergence of implicit approximation scheme. As mentioned above, 1-D explicit approximation scheme will be stable and convergent on the condition that γ is ranging from $0<\gamma\leqslant 1/2$. However, 1-D implicit approximation scheme is unconditional stable and convergent no mater what value γ is.

(4) Disadvantages. The choice of Δt is independent of Δx. However, for every time step a systems equation has to be solved.

3.4.2 2-D Transient Flow

In order to introduce the finite difference method for 2-D transient flow, we assume that the confined aquifer is homogeneous, isotropic, equal thickness and has no vertical recharge and discharge. The boundary conditions are first type and the vadose zone is rectangular.

1. *Conceptual hydrogeological model*

(1) Rectangular region.

(2) Homogeneous aquifer (T is constant).

(3) All boundaries are permeable.

(4) 2 - D transient flow.

2. *Mathematical model*

$$\begin{cases} T\left(\dfrac{\partial^2 H}{\partial x^2}+\dfrac{\partial^2 H}{\partial y^2}\right)=\mu\dfrac{\partial H}{\partial t} & \begin{matrix}0\leqslant x\leqslant a\\ 0\leqslant y\leqslant b\end{matrix} \quad t>0 \\ H|_{t=0}=H_0(x,y) & \begin{matrix}0\leqslant x\leqslant a\\ 0\leqslant y\leqslant b\end{matrix} \\ H|_{x=0}=\varphi_1(y,t) \quad H|_{x=a}=\varphi_2(y,t) & t>0 \\ H|_{y=0}=\psi_1(x,t) \quad H|_{y=b}=\psi_2(x,t) & t>0 \end{cases} \tag{3.65}$$

3. *Discretization*

Being the same to 1-D flow, we use two groups of lines parallel to the coordinate axis to divide the vadose region into mesh.

2-D space (Δx, Δy), $x_i=i\Delta x$ ($i=0$, 1, …, n)

$y_i=i\Delta y$ ($i=0$, 1, …, n)

And we divide the proposed calculation interval into several time steps, marked $t_n=n\Delta t$ ($n=0$, 1, …, m).

Our task is to obtain the heads $H(x_i, y_j, t_k)$ of each discrete node (x_i, y_i) at each time step t_k ($k=0$, 1, …, m) in vadose region. Generally, the spatial-temporal discrete nodes (x_i, y_j, t_k) are simplified as (i, j, k).

4. *Explicit approximation scheme*

(1) Establishment of equations. We assume that $H(x, y, t)$ is the exact solution to the model, for the node (i, j, k) there is a following equation (3.66):

$$T\left(\frac{\partial^2 H}{\partial x^2}+\frac{\partial^2 H}{\partial y^2}\right)_{i,j}^{k}=\mu\left.\frac{\partial H}{\partial t}\right|_{i,j}^{k} \tag{3.66}$$

We adopt the forward difference quotient to substitute for $\frac{\partial H}{\partial t}$, use second-order central-difference quotients to substitute for $\frac{\partial^2 H}{\partial x^2}$ and $\frac{\partial^2 H}{\partial y^2}$ respectively, that is:

$$\left.\frac{\partial H}{\partial t}\right|_{i,j}^{k}=\frac{h_{i,j}^{k+1}-h_{i,j}^{k}}{\Delta t} \tag{3.67}$$

$$\left.\frac{\partial^2 H}{\partial x^2}\right|_{i,j}^{k}=\frac{h_{i-1,j}^{k}-2h_{i,j}^{k}+h_{i+1,j}^{k}}{(\Delta x)^2} \tag{3.68}$$

$$\left.\frac{\partial^2 H}{\partial y^2}\right|_{i,j}^{k}=\frac{h_{i,j-1}^{k}-2h_{i,j}^{k}+h_{i,j+1}^{k}}{(\Delta y)^2} \tag{3.69}$$

Substituting the above three formulae into equation (3.66) and we can get:

$$T\left[\frac{h_{i-1,j}^{k}-2h_{i,j}^{k}+h_{i+1,j}^{k}}{(\Delta x)^2}+\frac{h_{i,j-1}^{k}-2h_{i,j}^{k}+h_{i,j+1}^{k}}{(\Delta y)^2}\right]=\mu\frac{h_{i,j}^{k+1}-h_{i,j}^{k}}{\Delta t}$$

Its truncation error is $O[(\Delta t)+(\Delta x)^2+(\Delta y)^2]$.

Let

$$\lambda_x=\frac{T\Delta t}{\mu(\Delta x)^2},\lambda_y=\frac{T\Delta t}{\mu(\Delta y)^2}$$

The above equation can be adapted as

$$h_{i,j}^{k+1}=\lambda_x h_{i-1,j}^{k}+\lambda_x h_{i+1,j}^{k}+\lambda_y h_{i,j+1}^{k}+\lambda_y h_{i,j+1}^{k}+(1-2\lambda_x-2\lambda_y)h_{i,j}^{k} \tag{3.70}$$
$$(i=1,2,\cdots,ni-1;j=1,2,\cdots,nj-1)$$

Thus, the equation relates to the water heads of six nodes, among which five are taken at time step n and only one is taken at time step $n+1$. If we know the values at time step n, each equation from equation (3.70) can be solved independently without simultaneous equations (Figure 3.8). Therefore, such a difference scheme is called two-dimensional explicit approximation scheme.

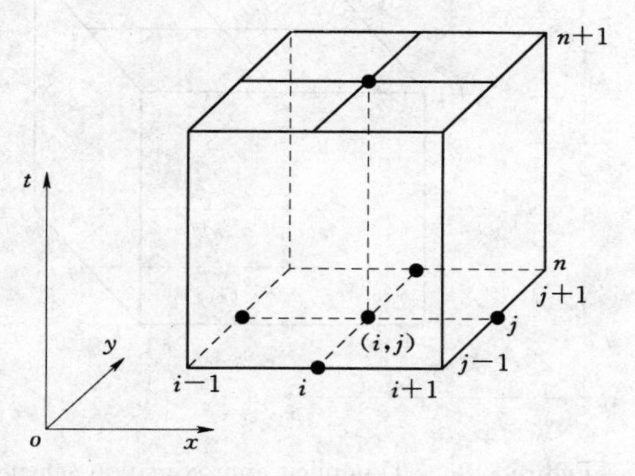

Figure 3.8 2-D explicit approximation scheme

(2) Solution procedures. We can see from equation (3.70) that if the water heads at time step k is known, the heads of nodes (i, j) $(i=1, 2, \cdots, ni-1;\ j=1, 2, \cdots, nj-1)$ can be solved directly at time step $k+1$. Then, using the heads of boundary nodes given by the boundary conditions in e-

quation (3. 65), the heads of all nodes at time step $k+1$ can be work out. When solution the equation, we assume $k=0$ and the heads at time t_1 can be calculated using the heads at time t_0 (initial time) by equation (3. 70). By analogy, we take $k=1$ and the heads at time t_2 can be determined by the heads at time t_1.

(3) Stability and convergence of Explicit Approximation Scheme. Similarly to 1-D explicit scheme, 2-D explicit approximation scheme is stable and convergent when $\lambda_x+\lambda_y\leqslant 1/2$.

5. *Implicit approximation scheme*

(1) Establishment of equations. The two-dimension spatial difference quotient of the implicit approximation scheme should be taken at time step $k+1$. As a result, the groundwater flow equation (3. 65) can be discretized into:

$$T\left[\frac{h_{i-1,j}^{k+1}-2h_{i,j}^{k+1}+h_{i+1,j}^{k+1}}{(\Delta x)^2}+\frac{h_{i,j-1}^{k+1}-2h_{i,j}^{k+1}+h_{i,j+1}^{k+1}}{(\Delta y)^2}\right]=\mu\frac{h_{i,j}^{k+1}-h_{i,j}^{k}}{\Delta t} \tag{3. 71}$$

Its truncation error is $O[(\Delta t)+(\Delta x)^2+(\Delta y)^2]$.

Let

$$\lambda_x=\frac{T\Delta t}{\mu(\Delta x)^2},\lambda_y=\frac{T\Delta t}{\mu(\Delta y)^2}$$

The above equation can be adapted as

$$-\lambda_x h_{i-1,j}^{k+1}-\lambda_x h_{i+1,j}^{k+1}-\lambda_y h_{i,j-1}^{k+1}-\lambda_y h_{i,j+1}^{k+1}+(1+2\lambda_x+2\lambda_y)h_{i,j}^{k+1}=h_{i,j}^{k} \tag{3. 72}$$
$$(i=1,2,\cdots,ni-1;j=1,2,\cdots,nj-1)$$

If we take space interval $\Delta x=\Delta y$, then $\lambda_x=\lambda_y=\lambda$, the above equation can be written as following:

$$-\lambda(h_{i-1,j}^{k+1}+h_{i+1,j}^{k+1}+h_{i,j-1}^{k+1}+h_{i,j+1}^{k+1})+(1+4\lambda)h_{i,j}^{k+1}=h_{i,j}^{k} \tag{3. 73}$$

Therefore, the equation contains the water head of one node at time step k and the water heads of five nodes at time step $k+1$ (Figure 3. 9). The solution can only be obtained by solving the simultaneous equations.

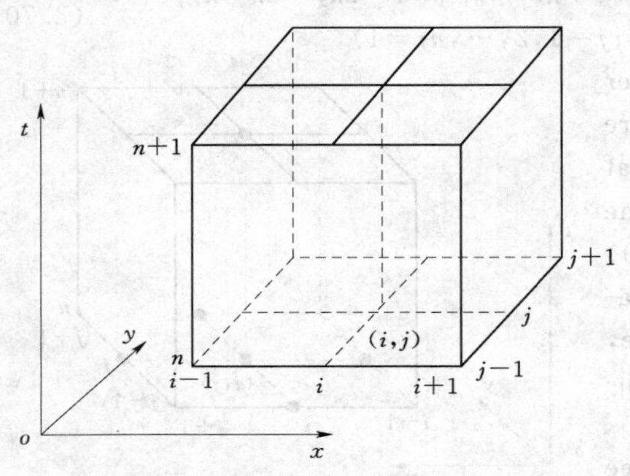

Figure 3. 9 2-D implicit approximation scheme

(2) Solution method (Iteration solutions). In order to calculate the water heads at time step $k+1$ by the heads at time step k, we take $i=1, 2, \cdots, ni-1$; $j=1, 2, \cdots, nj-1$, $(ni-1)\times(nj-1)$ equations can be established according to equation (3. 72). Under the first type boundaries, these equations just contain unknown heads of nodes at time step $k+1$ in vadose region. The number of the unknown values is $(ni-1)\times(nj-1)$, which is as the same as the number of equations. In each equation, the coefficient $(1+2\lambda_x+2\lambda_y)$ of $h_{i,j}^{k+1}$ must be placed on the main diagonal of the coefficient matrix when establishing equations. Only in this way can assure that the coefficient matrix

of algebraic equations is diagonally dominant strictly, and exist a unique solution.

The equation (3.72) can be adapted as

$$h_{i,j}=Wh_{i-1,j}+Eh_{i+1,j}+Nh_{i,j-1}+Sh_{i,j+1}+F \quad (i=1,2,\cdots,ni-1;j=1,2,\cdots,nj-1) \tag{3.74}$$

Where, the time step $k+1$ expressed by superscript of h are omitted, F represents the known items (including the heads at time step k). And if the node adjacent to (i, j) is located on the first type boundary, then it is also a known item and can be incorporated into item F.

1) Jacobi iteration. Taking initial value $h_{i,j}^{(0)}$, introducing it into the right side of equation (3.74) and obtaining the first improved values $h_{i,j}^{(1)}$, that is

$$h_{i,j}^{(1)}=Wh_{i-1,j}^{(0)}+Eh_{i+1,j}^{(0)}+Nh_{i,j-1}^{(0)}+Sh_{i,j+1}^{(0)}+F \quad (i=1,2,\cdots,ni-1;j=1,2,\cdots,nj-1)$$

Compute $h_{i,j}^{(2)}$ on the basis of $h_{i,j}^{(1)}$, we can get the improved values m^{th} with the continues repeating, and the $(m+1)$ times improved values can be computed according to the following equation, $h_{i,j}^{(m+1)}=Wh_{i-1,j}^{(m)}+Eh_{i+1,j}^{(m)}+Nh_{i,j-1}^{(m)}+Sh_{i,j+1}^{(m)}+F$.

If the absolute value of the difference between the adjacent improved values is less than a certain value at all nodes, the algorithm is convergent. So the following two inequalities can be used as iteration convergence criteria:

$$\max_{\substack{1\leqslant i\leqslant ni\\1\leqslant j\leqslant nj}}\{|h_{i,j}^{(m+1)}-h_{i,j}^{(m)}|\}<\varepsilon_1 \quad \text{or} \quad \max_{\substack{1\leqslant i\leqslant ni\\1\leqslant j\leqslant nj}}\left\{\frac{|h_{i,j}^{(m+1)}-h_{i,j}^{(m)}|}{|h_{i,j}^{(m+1)}|}\right\}<\varepsilon_2$$

2) Gauss—Siedel iteration. The advantage of Gauss—Siedel iteration is the utilization of the newly improved values at neighboring nodes in time. If the sequence of computation is row by row from top to bottom and node by node from left to right, then the iteration equation will be:

$$h_{i,j}^{(m+1)}=Wh_{i-1,j}^{(m+1)}+Eh_{i+1,j}^{(m)}+Nh_{i,j-1}^{(m+1)}+Sh_{i,j+1}^{(m)}+F \tag{3.75}$$

Its iteration convergence criterion is the same as that of Jacobi iteration.

3) Successive over-relaxation (SOR) method. Gauss—Siedel iteration formula can be adapted as follows:

$$h_{i,j}^{(m+1)}=h_{i,j}^{(m)}+\underbrace{[-h_{i,j}^{(m)}+Wh_{i-1,j}^{(m+1)}+Eh_{i+1,j}^{(m)}+Nh_{i,j-1}^{(m+1)}+Sh_{i,j+1}^{(m)}+F]}_{\Delta h_{i,j}^{(m+1)}}$$

The item enclosed in the parentheses of the right side of the equation can be considered as the correction value $\Delta h_{i,j}^{(m+1)}$ that caused by the iteration from the m times to the $m+1$ times, that is $h_{i,j}^{(m+1)}=h_{i,j}^{(m)}+\Delta h_{i,j}^{(m+1)}$.

In order to speed up the convergence rate, the correction value multiplied by ω is

$$\begin{aligned}h_{i,j}^{(m+1)}&=h_{i,j}^{(m)}+\omega[-h_{i,j}^{(m)}+Wh_{i-1,j}^{(m+1)}+Eh_{i+1,j}^{(m)}+Nh_{i,j-1}^{(m+1)}+Sh_{i,j+1}^{(m)}+F]\\&=\omega[Wh_{i-1,j}^{(m+1)}+Eh_{i+1,j}^{(m)}+Nh_{i,j-1}^{(m+1)}+Sh_{i,j+1}^{(m)}+F]+(1-\omega)h_{i,j}^{(m)}\end{aligned} \tag{3.76}$$

This is the successive over-relaxation (SOR) iterative formula. ω is called over-relaxation factor, ranging from 1 to 2.

So far, we have introduced three kinds of iteration methods for solving the linear simultaneous equation (3.75). Generally speaking, the convergence of iteration method for solving linear simultaneous equation is conditional. Based on the computing theory systems, the iteration method is convergent as long as the coefficient matrix of the simultaneous equation is diagonally dominant. Therefore, iteration methods for solving the implicit difference equations are convergent because the coefficient matrix of the equations is diagonally dominant.

Exercises

Solving the following definite problem. (2-D transient flow)

$$\begin{cases} T\left(\dfrac{\partial^2 H}{\partial x^2}+\dfrac{\partial^2 H}{\partial y^2}\right)=\mu\dfrac{\partial H}{\partial t} & \begin{matrix}0<x<800\\0<y<800\end{matrix} \quad t>0 \\ H|_{t=0}=0 & \begin{matrix}0<x<800\\0<y<800\end{matrix} \\ H|_{x=0}=1 \quad H|_{x=800}=0 \quad t>0 \\ \left.\dfrac{\partial H}{\partial y}\right|_{y=0}=\left.\dfrac{\partial H}{\partial y}\right|_{y=800}=0 \quad t>0 \\ T=20,\mu=0.002,\Delta x=\Delta y=100 \end{cases}$$

§ 3.5 Transient Flow in Unconfined Aquifers

The groundwater flow differential equation of a confined aquifer is linear because the thickness M of confined aquifer (or water conduction coefficient T) is irrelevant to head H. However, the thickness of an unconfined aquifer changes with the variation of water head, so the groundwater flow differential equation of an unconfined aquifer is nonlinear.

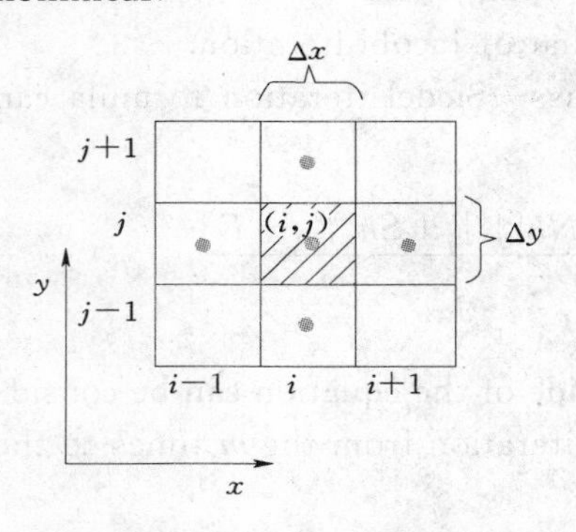

Figure 3.10 Equalization zone

Nonlinear partial differential equation can be solved analytically only for individual, extremely special conditions. For most cases, equations are solved by approximation methods after linerization.

For a 2-D unsteady groundwater flow in heterogeneous unconfined aquifer, we establish the differential equation of water balance in divergence form according to Darcy's law and the principle of water balance. So, taking the center of grid points (i, j) from the vadose zone as the equalization zone shown as Figure 3.10, taking the spatial step Δx, Δy respectively, time step Δt, we get

$$\begin{aligned} &T_{i-\frac{1}{2},j}\frac{h_{i-1,j}-h_{i,j}}{\Delta x}\Delta y+T_{i+\frac{1}{2},j}\frac{h_{i+1,j}-h_{i,j}}{\Delta x}\Delta y+T_{i,j-\frac{1}{2}}\frac{h_{i,j-1}-h_{i,j}}{\Delta y}\Delta x \\ &+T_{i,j+\frac{1}{2}}\frac{h_{i,j+1}-h_{i,j}}{\Delta y}\Delta x+Q_{ij}=\mu_{i,j}\frac{h_{i,j}^{n+1}-h_{i,j}^{n}}{\Delta t}\Delta x\Delta y \end{aligned} \tag{3.77}$$

Where: Q_{ij} is the water amount vetically flow into a grid per time.

In above equation, using the value of T in grid points (i, j) and the harmonic average T of the adjacent grid point to express $T_{i-\frac{1}{2},j}$、$T_{i+\frac{1}{2},j}$、$T_{i,j-\frac{1}{2}}$、$T_{i,j+\frac{1}{2}}$ respectively, and let

$$W_{i,j}=\frac{2T_{i-1,j}T_{i,j}}{T_{i-1,j}+T_{i,j}}\frac{\Delta y}{\Delta x}$$

$$E_{i,j}=\frac{2T_{i+1,j}T_{i,j}}{T_{i+1,j}+T_{i,j}}\frac{\Delta y}{\Delta x}$$

$$N_{i,j}=\frac{2T_{i,j-1}T_{i,j}}{T_{i,j-1}+T_{i,j}}\frac{\Delta x}{\Delta y}$$

$$S_{i,j}=\frac{2T_{i,j+1}T_{i,j}}{T_{i,j+1}+T_{i,j}}\frac{\Delta x}{\Delta y}$$

Where: $T_{i,j}=k_{i,j}(h_{i,j}-z_{i,j})$ is transmissibility coefficient in grid points $(i,\ j)$; $z_{i,j}$ is elevation of bottom aquifer in grid points $(i,\ j)$.

Then equation (3.77) becomes:

$$W_{i,j}h_{i-1,j}+E_{i,j}h_{i+1,j}+N_{i,j}h_{i,j-1}+S_{i,j}h_{i,j+1}+C_{i,j}h_{i,j}+Q_{i,j}=\mu_{i,j}\frac{\Delta x\Delta y}{\Delta t}(h_{i,j}^{n+1}-h_{i,j}^{n})\tag{3.78}$$

Where

$$C_{i,j}=-W_{i,j}-E_{i,j}-N_{i,j}-S_{i,j}$$

The equation (3.78) is the finite difference equation of unconfined groundwater.

We have not specified the head at which time step in the left equation. Obviously, the coefficients of the left equation contain the value of head, therefore, in accordance with its value at different time step, the difference equations may be linear or nonlinear. Here are several solutions.

1. *Explicit difference method*

We assume the head at time step n in the left equation (3.78) (including the head in coefficients), then we can get:

$$h_{i,j}^{n+1}=\frac{\Delta t}{\mu_{i,j}\Delta x\Delta y}(W_{i,j}^{n}h_{i-1,j}^{n}+E_{i,j}^{n}h_{i+1,j}^{n}+N_{i,j}^{n}h_{i,j-1}^{n}+S_{i,j}^{n}h_{i,j+1}^{n}+\widetilde{C}_{i,j}^{n}h_{i,j}^{n}+Q_{i,j})\tag{3.79}$$

Where

$$\widetilde{C}_{i,j}^{n}=C_{i,j}^{n}+\frac{\Delta x\Delta y}{\Delta t}\mu_{i,j}$$

Apparently, it is the most simple method. It can calculate the head at time step $n+1$ directly from the head at time step n without solving simultaneous equations. But its accuracy is poor and is conditionally stable.

2. *Explicit-implicit difference method*

We assume the head in coefficients (description of flow thickness) of left equation (3.78) are explicit, namely taking the head value at time step n, while assume the head describing the hydraulic gradient are implicit, namely taking the head value at time step $n+1$, then we can get

$$W_{i,j}^{n}h_{i-1,j}^{n+1}+E_{i,j}^{n}h_{i+1,j}^{n+1}+N_{i,j}^{n}h_{i,j-1}^{n+1}+S_{i,j}^{n}h_{i,j+1}^{n+1}+C_{i,j}^{n}h_{i,j}^{n+1}=F_{i,j}\tag{3.80}$$

Where

$$C_{i,j}^{n}=-W_{i,j}^{n}-E_{i,j}^{n}-N_{i,j}^{n}-S_{i,j}^{n}-\frac{\Delta x\Delta y}{\Delta t}\mu_{i,j}$$

$$F_{i,j}=-\frac{\Delta x\Delta y}{\Delta t}\mu_{i,j}h^{n}_{i,j}-Q_{i,j}$$

At this time, equation is linear and can be solved according to the general solution.

This method is applicable to the aquifer thickness is large, and the head changed is small in a period of time Δt. Generally speaking, $\frac{\Delta h}{h-z}\leqslant 5\%$.

3. *Implicit difference method*

This method is also known as the fully implicit method. Taking the head value at time step $n+1$ (including the heads in coefficients) in the left equation (3. 78), then we can get

$$W^{n+1}_{i,j}h^{n+1}_{i-1,j}+E^{n+1}_{i,j}h^{n+1}_{i+1,j}+N^{n+1}_{i,j}h^{n+1}_{i,j-1}+S^{n+1}_{i,j}h^{n+1}_{i,j+1}+C^{n+1}_{i,j}h^{n+1}_{i,j}=F_{i,j} \quad (3.81)$$

Where

$$C^{n+1}_{i,j}=-W^{n+1}_{i,j}-E^{n+1}_{i,j}-N^{n+1}_{i,j}-S^{n+1}_{i,j}-\mu_{i,j}\frac{\Delta x\Delta y}{\Delta t}$$

$$F_{i,j}=-\frac{\Delta x\Delta y}{\Delta t}\mu_{i,j}h^{n}_{i,j}-Q_{i,j}$$

This is a nonlinear equation, which is more complex to solve. If we predefine the coefficients (given the head value in coefficient), then the equation is linear. Therefore, the equation usually solved by iterative method.

For the sake of convenience, we will temporarily omit the left identified time step $n+1$, so equation (3. 81) can be written as

$$W_{i,j}h_{i-1,j}+E_{i,j}h_{i+1,j}+N_{i,j}h_{i,j-1}+N_{i,j}h_{i,j+1}+C_{i,j}h_{i,j}=F_{i,j} \quad (3.82)$$

In order to solve the head value at time step $n+1$ by time step n, we should conduct iterative calculation as the following steps.

Step 1 Taking a group of head value $h^{(0)}$ as initial iterative coefficient value of calculation equation (usually taking the head value at time step n), the equation becomes a linear equation after calculation of the coefficients and then solve the head denoted as $h^{(1)}$ according to solution of linear equations solving. Taking $h^{(1)}$ as the approximate head value at time step $n+1$.

Step 2 Taking $h^{(1)}$ as the head value of coefficient in calculation equation and calculate the coefficients, then can work out the head denoted as $h^{(2)}$ by the method of a set of linear algebraic equations solving. Taking $h^{(2)}$ as the second approximate head value at time step $n+1$.

Repeated iterative calculation after several times (such as m times), If the allowable error ε_1 and ε_2 are given in advance, and adjacent two iterative solution $h^{(m+1)}$ and $h^{(m)}$ meets following equations,

$$\max_{\substack{1\leqslant i\leqslant Ni\\ i\leqslant j\leqslant Nj}}\{|h^{(m+1)}_{i,j}-h^{(m)}_{i,j}|\}<\varepsilon_1 \text{ and } \max_{\substack{1\leqslant i\leqslant Ni\\ 1\leqslant j\leqslant Nj}}\left\{\frac{|h^{(m+1)}_{i,j}-h^{(m)}_{i,j}|}{|h^{(m+1)}_{i,j}|}\right\}<\varepsilon_2$$

Then, $h^{(m)}_{i,j}$ can be seen as the head value at time step $n+1$ at grid points (i, j). Thus, we completed the calculation from time step n to $n+1$.

If we solve the linear equation by iterative method at each iteration, then the whole solving process is double iteration.

Despite of large amount of calculation, implicit difference method has high precision, stability and good convergence for solving unconfined flow problems.

Exercise

To solve following 2-D transient flow in an unconfined aquifer.

$$\begin{cases} \dfrac{\partial}{\partial x}\left(K(H-Z)\dfrac{\partial H}{\partial x}\right)+\dfrac{\partial}{\partial y}\left(K(H-Z)\dfrac{\partial H}{\partial y}\right)=\mu\dfrac{\partial H}{\partial t} & \begin{matrix} 0<x<1000 \\ 0<y<1000 \end{matrix} \quad t>0 \\ H|_{t=0}=0 & \\ H|_{B_1}=1 \qquad t>0 & \\ K=5, \mu=0.2, Z=2 & \end{cases}$$

$B_1(0<x<1000$, $0<y<1000)$ are the boundaries of rectangle (All four boundaries are the first type boundary). Taking $\Delta x=100$, $\Delta y=100$, Δt is determined by yourself. (Solving by Explicit difference method and Explicit-implicit difference method respectively.)

Chapter 4 Introduction and Tutorial of Visual MODFLOW

§ 4.1 Introduction of Visual MODFLOW

4.1.1 Brief Introduction of Visual MODFLOW

Visual MODFLOW is the most complete, and user-friendly, modeling environment for practical applications in three-dimensional groundwater flow and contaminant transport simulation. This fully-integrated package combines powerful analytical tools with a logical menu structure. Easy-to-use graphical tools allow you to:

(1) Quickly mark the model domain and select units.

(2) Conveniently assign model properties and boundary conditions.

(3) Run model simulations for flow and contaminant transport.

(4) Calibrate the model using manual or automated techniques.

(5) Optimize pumping and remediation well rates and locations.

(6) Visualize the results using 2D or 3D graphics.

The model input parameters and results can be visualized in 2D (cross-section and plan view) or 3D at any time during the development of the model or the displaying of the results. For complete three-dimensional groundwater flow and contaminant transport modeling, Visual MODFLOW is the best software package available.

4.1.2 About the Interface

The Visual MODFLOW interface has been specifically designed to increase modeling productivity, and decrease the complexities of three-dimensional groundwater flow and contaminant transport models. In order to simplify the graphical interface, Visual MODFLOW is divided into three separate modules:

(1) The Input section.

(2) The Run section.

(3) The Output section.

Each of these modules is accessed from the Main Menu when a Visual MODFLOW project is started, or an existing project is opened. The Main Menu serves as a link to seamlessly switch between each of these modules, to define or modify the model input parameters, run the simulations, calibrate and optimize the model, and display results.

(1) The Input section. The Input section allows the user to graphically assign all of the necessary input parameters to build a three-dimensional groundwater flow and contaminant transport model. The input menus have the basic "model building blocks" for

assembling a data set using MODFLOW, MODFLOW-SURFACT, MODPATH, Zone Budget, and MT3D/RT3D. These menus are displayed in a logical order to guide the modeler through the steps necessary to design a groundwater flow and contaminant transport model.

(2) The Run section. The Run section allows the user to modify the various MODFLOW, MODFLOW-SURFACT, MODPATH, MT3D/RT3D, MGO, and WinPEST parameters and options which are run-specific. These include selecting initial head estimates, setting solver parameters, activating the re-wetting package, specifying the output controls, etc. Each of these menu selections has default settings, which are capable of running most simulations.

(3) The Output section. The Output section allows the user to display all of the modeling and calibration results for MODFLOW, MODFLOW-SURFACT, MODPATH, Zone Budget, and MT3D/RT3D. The output menus allow you to select, customize, overlay the various display options, and export images to present the modeling results.

4. 1. 3 Main Menu Screen

The Main Menu screen contains the following options:

File: Select a file utility, select import and export features, or exit Visual MODFLOW.

Input: Design, modify, and visualize the model grid and input parameters.

Run: Enter or modify the model run settings, and run the numerical simulations in either project mode or batch mode.

Output: Visualize the model output (simulation results) using contour maps, color maps, velocity vectors, path-lines, time-series graphs, scatter plots and so on.

Setup: Select the "Numeric Engine" for both the flow and transport simulation.

Help: Get general information on how to use Visual MODFLOW.

4. 1. 4 Screen Layout

After opening or creating a Visual MODFLOW project and selecting the Input module, a screen layout similar to Figure 4. 1 will appear.

(1) Top menu bar. This bar contains specific data category menus for each section of the interface (Input, Run, and Output).

(2) Side menu bar. This bar contains the View Control buttons plus tool buttons specific for each data category. The view options are as follows:

[View Column]: View a cross-section along a column.

[View Row]: View a cross-section along a row.

[View Layer]: Switch from cross-section to planview.

[Goto]: View a specified column, row or layer.

[Previous]: View previous column, row or layer.

[Next]: View next column, row, or layer.

(3) Navigator cube. Providing a simplified 3-D representation of the model domain highlighting the current layer, column, and row.

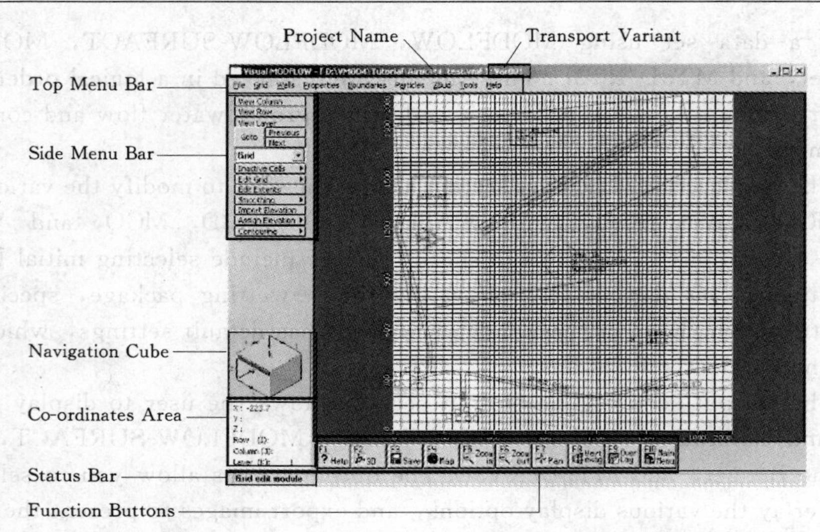

Figure 4. 1 Screen layout of Visual MODFLOW

(4) Co-ordinates area. When in plan view of a given layer, it shows the current location of the cursor; X and Y represent model co-ordinates, and Z represents each grid cell center at which the head is calculated. The value of i, j, k are the current cell indices on which the cursor is located.

(5) Status bar. The bar describes the function and use of the feature currently selected or pointed to by the cursor.

(6) Function buttons. Common functions to the Input, Run, and Output screens can be selected by clicking on the button or by pressing the function key on your keyboard. The function buttons are:

F1: Open the general help window.

F2: Open the Visual MODFLOW 3D - Explorer.

F3: Save the current data for the model.

F4: Import and display. DXF, SHP, or graphics image files for use as a site map.

F5: Zoom in a selected rectangular area.

F6: Reset the display area to show the entire model domain.

F7: Drag the current view of the model in any direction to display a different model region. Double-click in the model domain to refresh the screen.

F8: Set the vertical exaggeration value for viewing cross-sections along a row or column.

F9: Opens the Overlay Control window to make selected overlays visible and to customize the priority (order) in which they appear.

F10: Returns to the Main Menu screen.

§ 4. 2 Instructions of Example Model

1. *About the Visual MODFLOW tutorial*

This following section will present a tutorial, with a complete set of input and output files, for an example model (Airport) that will allow you to examine the post-processing features and capabilities of Visual MODFLOW4. 2. The numerical simulations for this problem have already been completed to allow you to evaluate the output visualization features for the sample model results.

This tutorial guides you through some of the steps necessary to design and run a model, and visualize the results. The instructions for the tutorial are provided in a stepwise format that allows you to choose the features that you are interested in examining without having to complete the entire exercise.

2. *Description of the example model*

The site is located near an airport just outside of Waterloo. The surficial geology at the site consists of upper sand and gravel aquifer, lower sand and gravel aquifer, and a clay and silt aquitard separating the upper and lower aquifers. The relevant site features consist of a plane refueling area, a municipal water supply well field, and a discontinuous aquitard zone. These features are illustrated in Figure 4. 2.

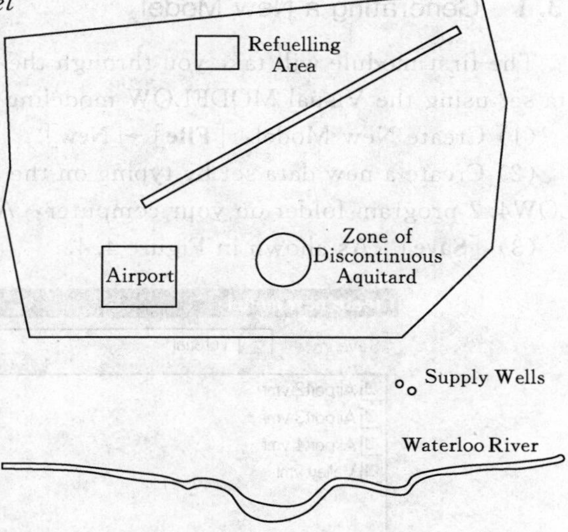

Figure 4. 2 Description of the model zone

The municipal well field consists of two wells. The east well pumps at a constant rate of 550m^3/d, while the west well pumps at a constant rate of 400m^3/d. Over the past ten years, airplane fuel has periodically been spilled in the refueling area and natural infiltration has produced a plume of contamination in the upper aquifer. This tutorial will guide you through the steps necessary to build a groundwater flow model for this site. This model will demonstrate the potential impact of the fuel contamination on the municipal water supply wells.

When discussing the site, in plan view, the top of the site will be designated as north, the bottom of the site as south, the left as west, and the right as east.

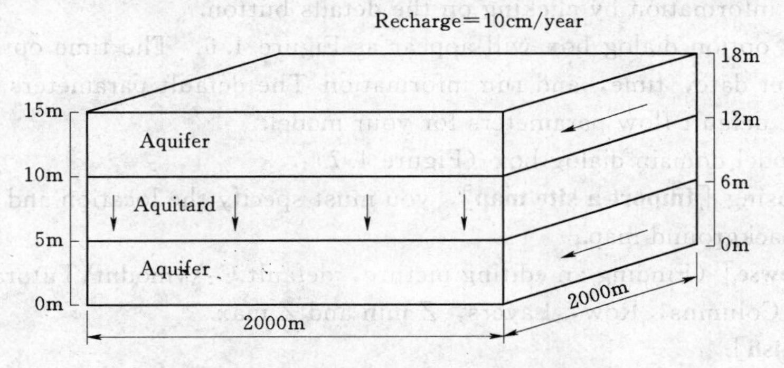

Figure 4. 3 Spatial features of the model zone

Groundwater flow is from north to south (top to bottom) in a three-layer system consisting of an upper unconfined aquifer, an intervening middle aquitard, and a lower confined aquifer, as illustrated in the following figure. The upper aquifer and lower aquifers have hydraulic conductivities of $2e^{-4}$ m/sec and the aquitard has a hydraulic conductivity of $1e^{-10}$ m/sec. These features are shown in Figure 4.3.

§ 4.3 Creating and Defining a Flow Model

4.3.1 Generating a New Model

The first module will take you through the steps necessary to generate a new model data set using the Visual MODFLOW modeling environment.

(1) Create New Model: [**File**]→[**New**].

(2) Create a new data set by typing on the Tutorial subfolder of your Visual MODFLOW4.2 program folder on your computer: Airport.

(3) [**Save**]. As shown in Figure 4.4.

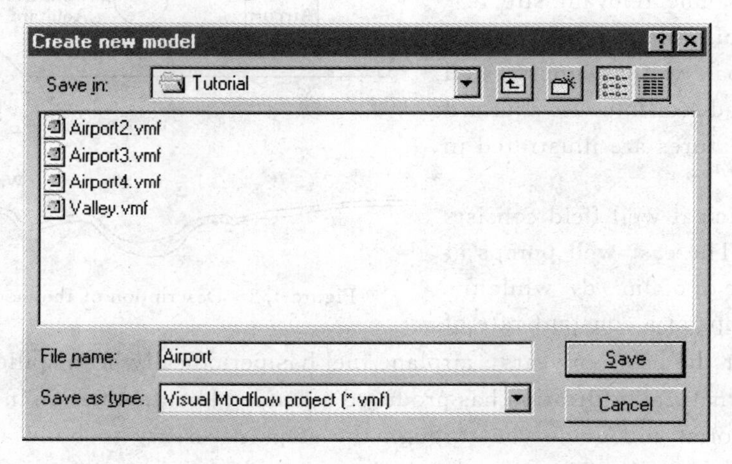

Figure 4.4 Creating a new model

The project information dialog box will appear as Figure 4.5. The project information frame allows you to enter an optional project title, project description, unit and the flow type information by clicking on the details button.

The flow option dialog box will appear as Figure 4.6. The time option frame allows you to set date, time, and run information. The default parameters frame allows you to set the default flow parameters for your model.

In the model domain dialog box (Figure 4.7):

(1) Choosing [**Import a site map**], you must specify the location and the file name of the. DXF background map.

(2) [**Browse**] (Finding an editing picture, default C:\vmodnt\Tutorail).

(3) Edit Columns, Row, Layers, Z min and Z max.

(4) [**Finish**].

A dialog box will appear prompting you to define the extents of the model area

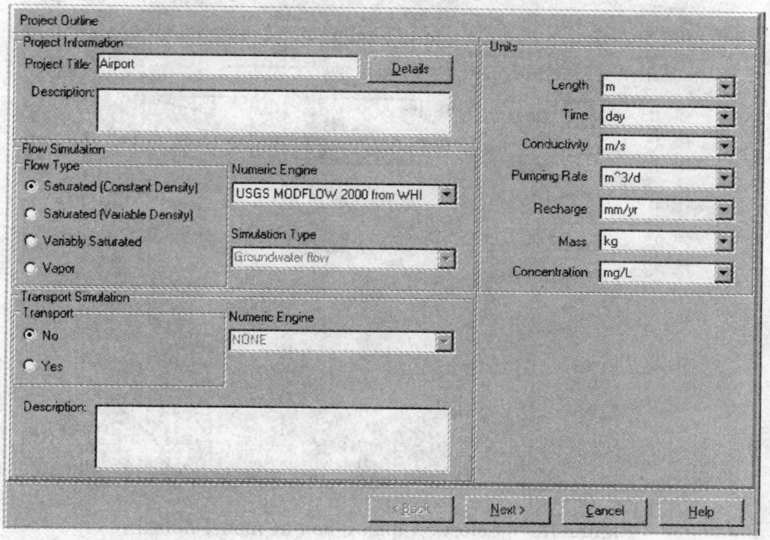

Figure 4. 5 Selection dialog box of the project outline

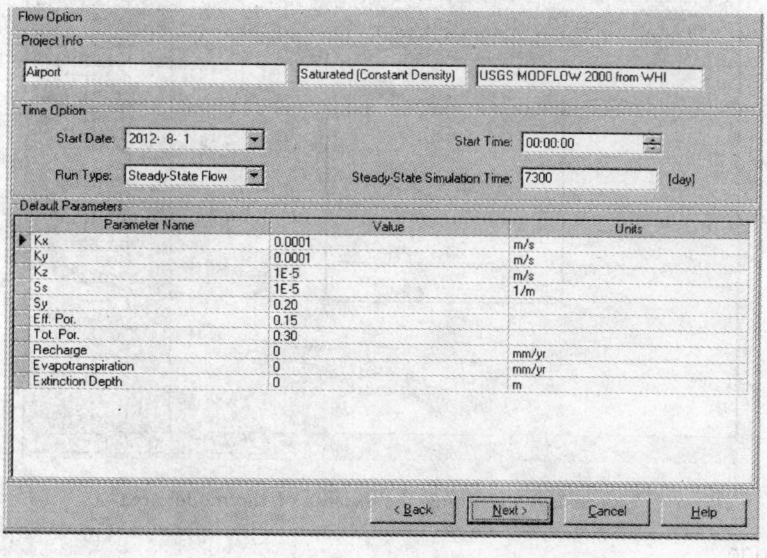

Figure 4. 6 Selection dialog box of project information

(Figure 4. 8). Visual MODFLOW will read the minimum and maximum coordinates from the site map and suggest a default that is centered in the model domain. Visual MODFLOW now allows the user to rotate and align the model grid over the site map, use local co-ordinates, and set the extents of the DXF map. If a bitmap was used as background map, the image can be geo-referenced and the grid can be aligned on the bitmap as well. Type the following values over the numbers shown on the screen:

Display Area

$X1 \rightarrow 0$

$Y1 \rightarrow 0$

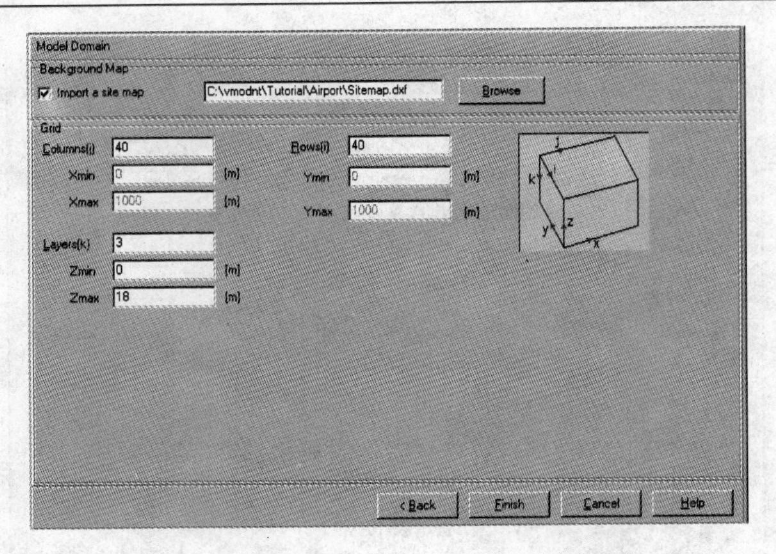

Figure 4. 7 Selection dialog box of model domain

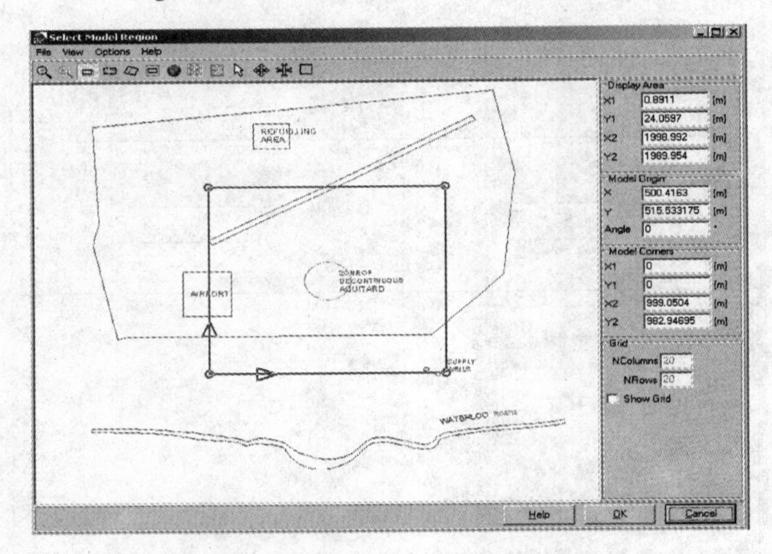

Figure 4. 8 Defining the extents of the model area

$X2$ →2000
$Y2$ →2000
Model Origin
X →0
Y →0
Angle →0
Model Corners
$X1$ →0
$Y1$ →0
$X2$ →2000
$Y2$ →2000

Clicking [**OK**].

A uniformly spaced finite difference grid will be automatically generated within the model domain and a site base map will appear on the screen as shown in Figure 4. 9.

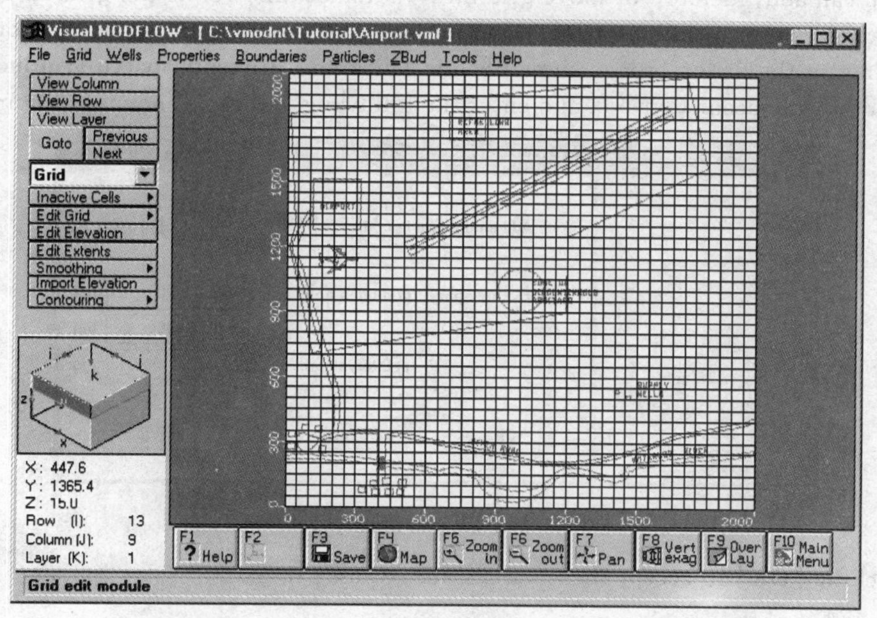

Figure 4. 9 Uniformly spaced finite difference grid

This is the Visual MODFLOW Input module. The top menu bar of the Input module is divided into the primary "building blocks" for any groundwater flow and contaminant transport model (Grid, Wells, Properties, Boundaries, and Particles). Each of these menu items provides access to an associated Input screen. The left-hand menu bar provides a list of graphical tools associated with the selected Input screen. The Navigator Cube on the lower left-hand side of the screen shows the location of the current row, layer, or column in the model, and the *X*, *Y*, *Z*, and *I*, *J*, *K* location of the mouse pointer.

4. 3. 2 Refining the Model Grid

This section will refine the model grid in the areas of interest such as around the supply wells and around the refueling area. The reason for refining the grid is to get more detailed simulation results in area of interest or in zones where you anticipate steep hydraulic gradients. If drawdown is occurring around the well, the water level will have a smoother surface with finer grid spacing.

You are now located in the input module where the "building blocks" for a model appear as items in the menu bar. Visual MODFLOW loads the [Grid] input screen when you first enter the input module. The top six toolbar buttons on the left-hand side of the screen ([View Column], [View Row], [View Layer], [Go to], [Previous], and [Next]) appear on every screen and allow you to change the model display from plan view to cross-section at any time. The remaining toolbar buttons describe the various functions that can be performed to modify the model grid.

1. *To refine the grid in the x-direction*

Clicking [**Edit Grid**]→[**Edit Columns**].

You can add, delete, or move gridlines, automatically refine the grid. In addition, you can add vertical gridlines at specified intervals by clicking on the "right mouse button" bring up the dialog box as shown below. Clicking "⊙**Evenly spaced gridlines from**" and inside the adjacent text box and enter the following values as shown in Figure 4. 10.

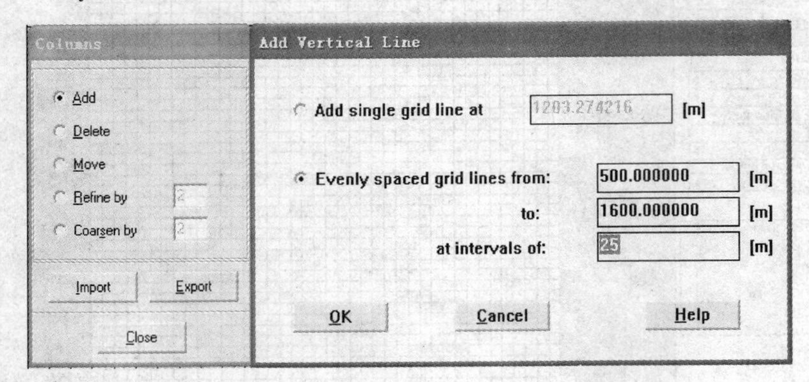

Figure 4. 10 Refining the model grid

Clicking [**OK**] to accept these values and [**Close**] to exit.

2. *To refine the grid in the y-direction*

Now we refine the grid in the y-direction from the refueling area to the supply wells. The method is same as x-direction.

[**Edit Grid**]→[**Edit Rows**].

Click "⊙**Evenly spaced gridlines from**" and enter the following values: from 400 to 1900, at intervals of 25. The refined grid should appear as shown in Figure 4. 11.

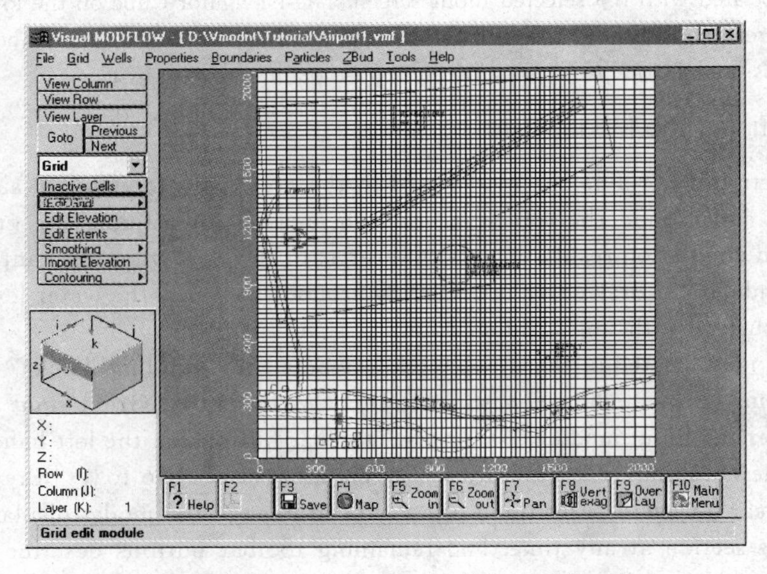

Figure 4. 11 Refining result of the model grid

3. *To view the model in cross-section*

Clicking [**View Column**] from the left-hand menu.

Move the mouse cursor to anywhere in the grid. As you move the cursor across the screen, a bar will highlight the column according to the cursor location. To select a column to view, click the left mouse button on the desired column. Visual MODFLOW transfers the screen display from plan view to a cross-sectional view of the model grid. At this point the model has no vertical exaggeration and the cross-section will appear as a thick line with the three layers barely discernible. To properly display the three layers you will need to add vertical exaggeration to the cross-section.

Clicking [**F8 – Vert Exag**] from the bottom of the screen. A Vertical Exaggeration dialog box (Figure 4. 12) appears prompting you for a vertical exaggeration value.

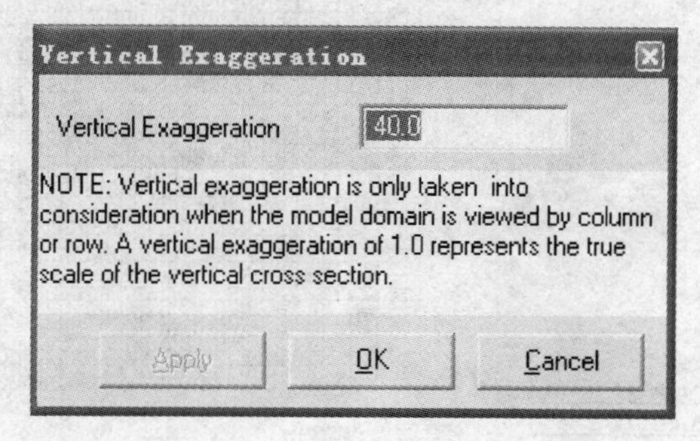

Figure 4. 12 Vertical exaggeration dialog box

Type: 40 and press [**OK**].

The three layers of the model will then be displayed on the screen as shown below. From Figure 4. 13 you can see that each layer has a uniform thickness across the entire cross-section.

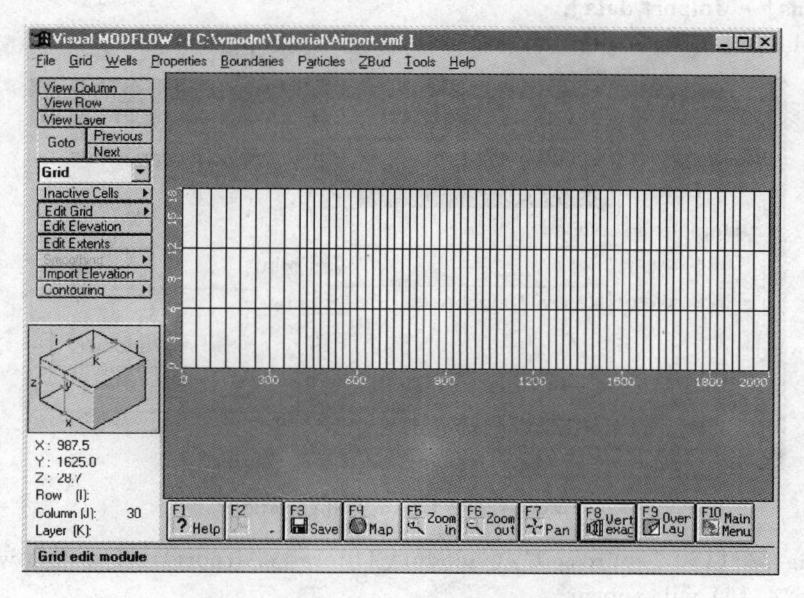

Figure 4. 13 Cross-section of the model

4. *Import variable layer elevations*

Visual MODFLOW allows you to import variable layer elevations from surfer. grd files or from space delimited x, y, z ASCII files. In this example, we will import an ASCII x, y, and z file to create a sloping ground surface topography and layers with variable thickness. Press [**Import Elevation**] from the left-hand menu and the following dialog box (Figure 4. 14) will appear:

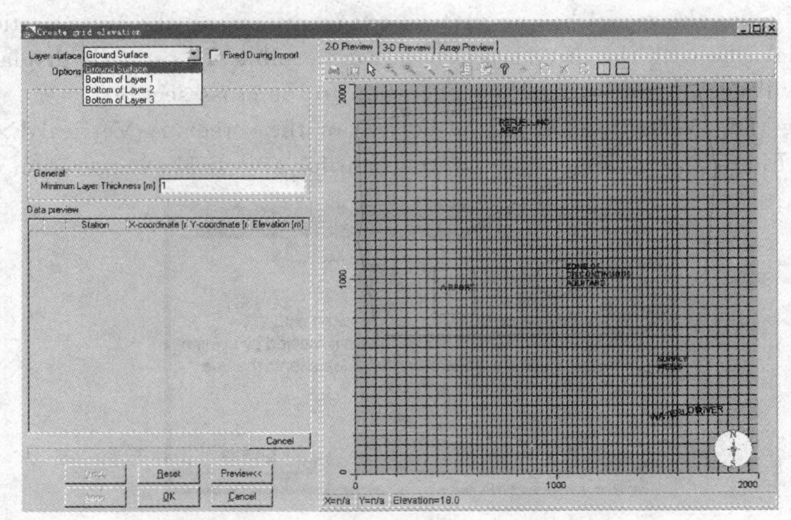

Figure 4. 14 Dialog box for elevation importing

Then it will be successively imported elevations of Groundwater surface, Bottom layer 1, Bottom layer 2 and Bottom layer 3. First, we import the elevations for Ground surface.

[**Options**]→[**Import data**].

Then the following dialog box (Figure 4. 15) will appear.

Figure 4. 15 Create grid elevation

Clicking (Data source, C:\vmodnt\Tutorail\Airport). The following dialog box (Figure 4. 16) will appear.

Figure 4. 16 Selecting the file of surface elevation

Clicking [**Open**], then the dialog box is appearing (Figure 4. 17). Modify "Match to column number" to connect the data with their corresponding data base. Clicking [**Next**], the dialog box is shown as Figure 4. 18.

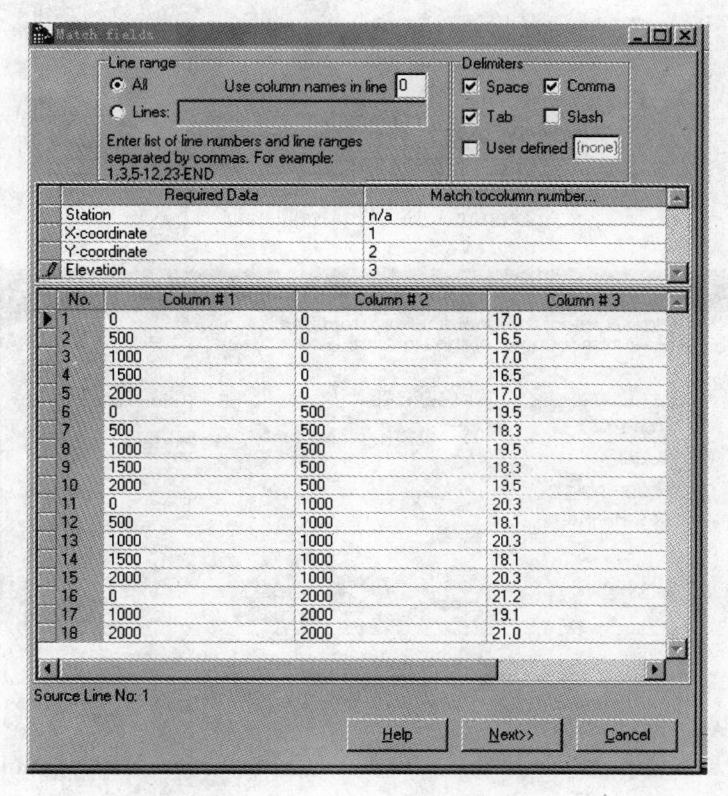

Figure 4. 17 Match fields

The error data will be shown by red color, correcting it and then clicking [**Finish**]. Then the dialog box is shown as Figure 4. 19.

Clicking [**OK**], the dialog box is shown as Figure 4. 20.

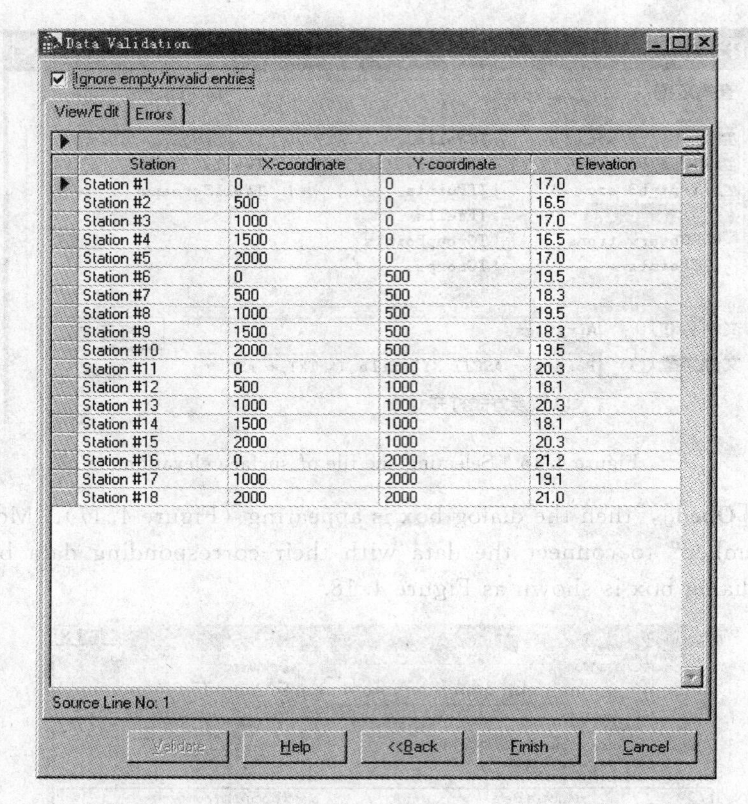

Figure 4. 18 Data validation

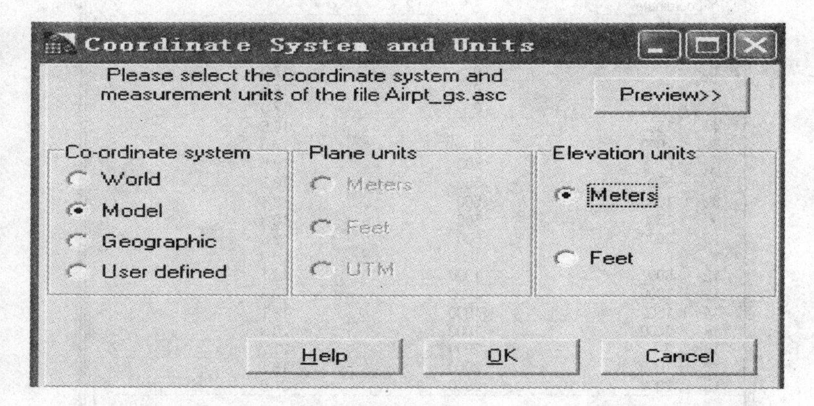

Figure 4. 19 Selecting the coordination system and units

Clicking [**Apply**] and [**OK**].

Using the same procedures, the elevation of other layers will be imported respectively. Finally, Figure 4. 21 is shown.

5. *Divide layer*

[**Edit Grid**]→[**Edit layers**].

Move the mouse into the model cross-section. A deforming horizontal line will fol-

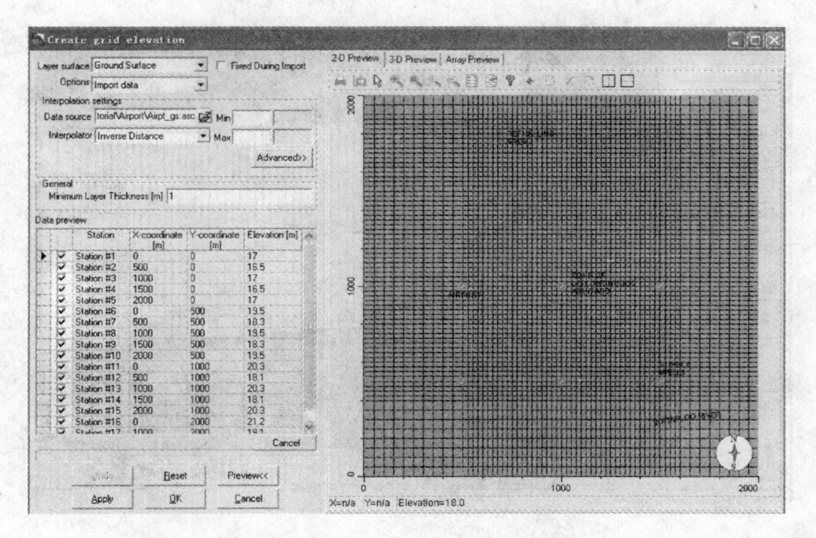

Figure 4. 20 Creating grid elevation

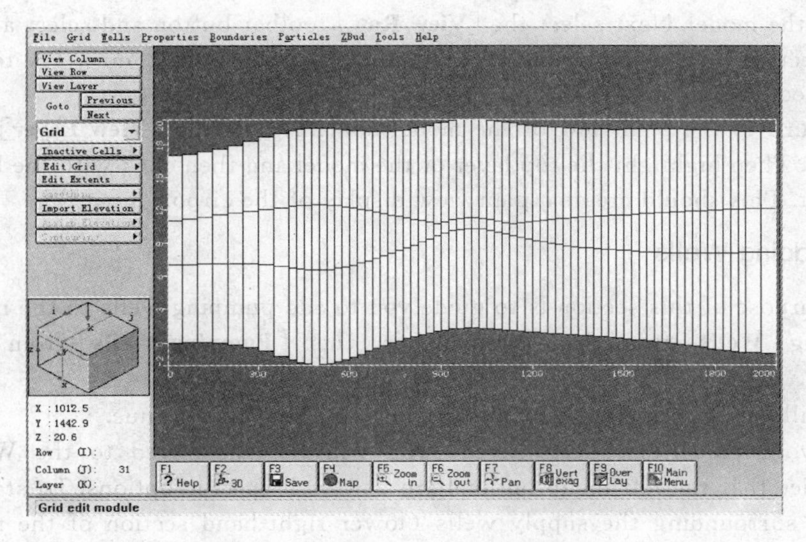

Figure 4. 21 Elevation of the 3 layers

low the mouse elevation within the model cross-section. To sub-divide a layer, approximate vertical mid-point of layer 1 and press the left mouse button. A layer division will be added in this location.

Or move the mouse and the horizontal line is within the layer then press the **RIGHT MOUSE BUTTON.** An **Add Horizontal Layer** dialog box will appear. Clicking "⊙**Split current layer into '2' evenly spaced layers**" and [**OK**].

Repeat this action for the bottom layer of the model.

When you have completed these instructions, Clicking [**Close**] to exit.

The model cross-section should now consist of six layers and should appear similar to Figure 4. 22.

Use the [**Next**] and [**Previous**] buttons on the left toolbar to view the column cross-

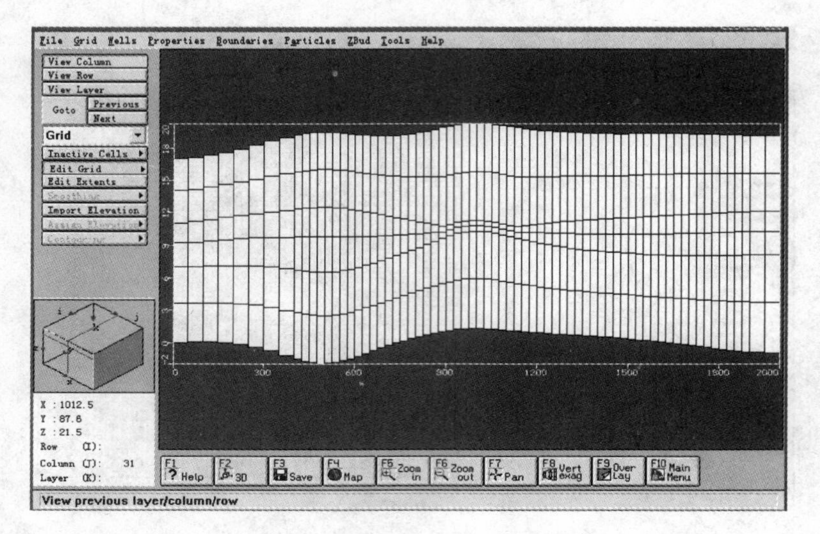

Figure 4. 22 Result of the layer splitting

sections of the model. Next select the [**View Row**] toolbar button and select a row from the cross-section by clicking on one of the highlighted vertical columns. Try to view the row cross-sections of the model.

To return to the plan-view of the model domain, clicking [**View Layer**] from the left toolbar. Then highlight the top layer of the model and then click with the left mouse button on it. This should create a plan view display of the airport site.

4. 3. 3 Adding Wells

The purpose of this section is to guide you to add pumping wells to the model.

Clicking [**Wells**] from the top menu bar, then [**Pumping Wells**] from the drop-down menu.

You will be asked to save your data, clicking [**Yes**] to continue.

Once your model has been saved, you will be transferred to the Well Input screen. Notice that the left menu options are new well-specific options. To start, zoom in the area surrounding the supply wells (lower right-hand section of the model domain).

Clicking [**F5-Zoom In**].

Move the cursor to the upper left of the supply wells and click on the left mouse button. Then stretch a box around the supply wells and click again to close the zoom window.

1. *To add a pumping well to your model*

Clicking [**Add Well**].

Move the cursor to the west well and click on it. A new well dialog box (Figure 4. 23) will appear prompting you for specific well information.

Inputting **Well Name**: Supply Well 1.

To add a well screen interval, we click the column labeled **"Screen Bottom"** and enter the values.

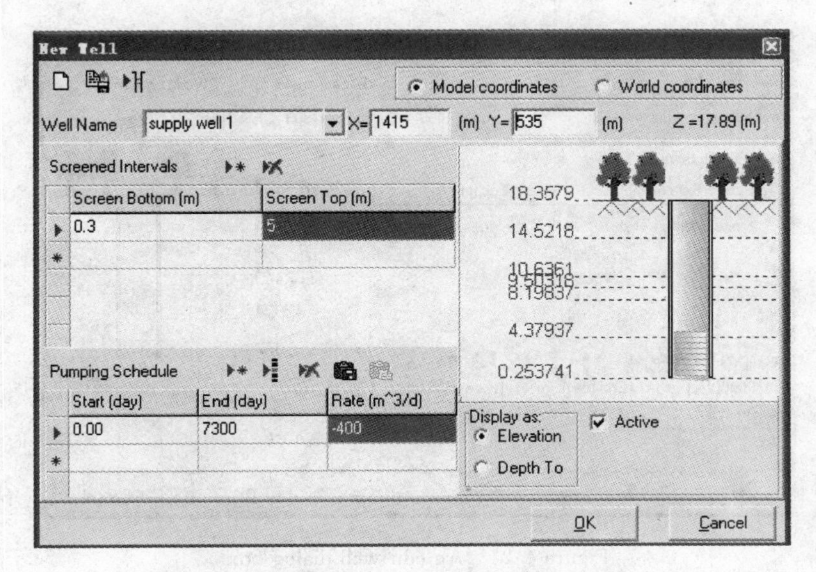

Figure 4. 23 Well information

Notice the well screen interval appears in the well bore on the right-hand side of the window. You can graphically assign the well screen by clicking the right mouse button on the well bore. Then click and drag the mouse where you want to add the well screen.

To enter the well pumping schedule, we click the left mouse button once inside the text box under the column labeled **"End (day)"** and enter the information.

Clicking [**OK**] to accept this well information.

If you have failed to enter any required data Visual MODFLOW will prompt you to complete the table at this time.

2. *Copy a pumping well to your model*

This step is to assign the well parameters for the second pumping well by using a short-cut which allows you to copy the characteristics from one well to another location.

Clicking [**Copy Well**] from the left toolbar.

Move the cursor on top of the west well and click on it. Then move the cursor to the location of the east well location and click again to copy the well. Next you will edit the well information from the copied well.

Clicking [**Edit Well**] from the left toolbar.

Select the new east well by clicking on the well symbol. An edit well dialog box (Figure 4. 24) will appear:

We input some values as figure shown.

Clicking [**OK**] to accept these well parameters, then [**F6-Zoom Out**].

4. 3. 4 Assigning Model Properties

1. *Assign hydraulic conductivities*

Clicking [**Properties**]→[**Conductivity**], then clicking [**Yes**] to continue.

Once your model has been saved, you will be transferred to the model properties

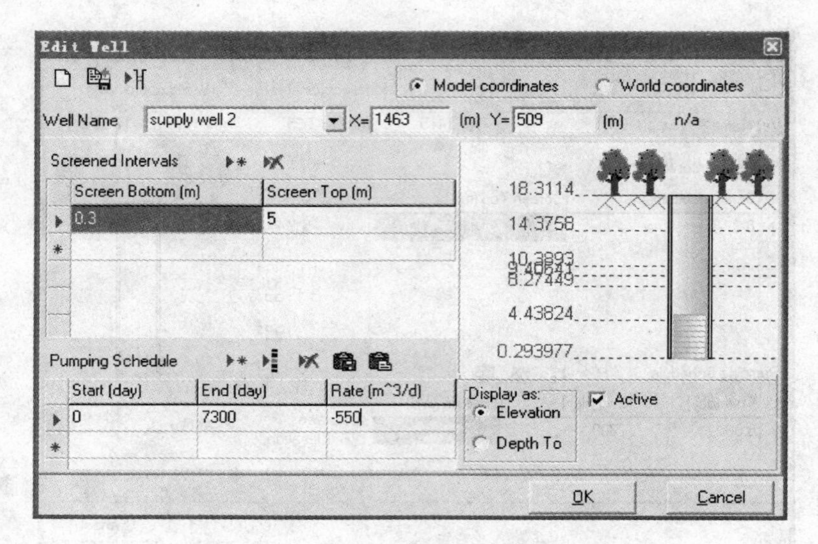

Figure 4. 24 An edit well dialog box

Input screen.

Clicking [**Databases**], a dialog box will appear as Figure 4. 25.

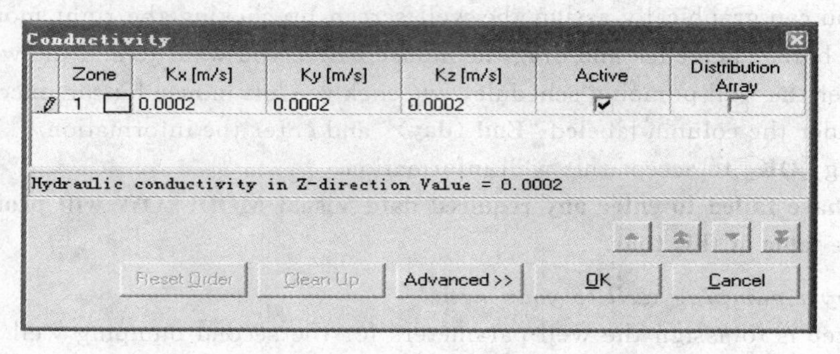

Figure 4. 25 Database of the conductivity

Modify the default values of hydraulic conductivity and click [**OK**] to accept values.

Then we will assign conductivity values to the aquitard that contain layers 3 and 4. Therefore, we should go to layer 3 by clicking [**Go to**] from the left toolbar (Figure 4. 26) .

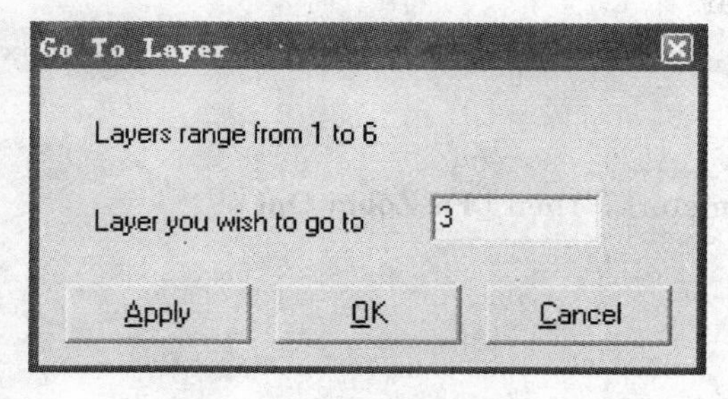

Figure 4. 26 Go to layer

Clicking [**OK**].

We are now viewing the third layer of model. In this six layer model, layers 3 and 4 represent the aquitard separating the upper and lower aquifers. The next step is to assign a lower hydraulic conductivity value to layers 3 and 4.

Clicking [**Assign**]→ [**Window**].

Move the mouse to the north-west corner of the grid and click on the centre of the cell. Then move the mouse to the southeast corner and click on the centre of the cell. This creates a window covering the entire layer. An assign K property dialog box will appear (Figure 4. 27).

Conductivity - [Assign Window]

Zone # 2 New

Name	Value	Units	Description
Kx	1e-10	[m/s]	Hydraulic conductivity in X-direction
Ky	1e-10	[m/s]	Hydraulic conductivity in Y-direction
Kz	1e-11	[m/s]	Hydraulic conductivity in Z-direction

OK Cancel

Figure 4. 27 Assigning conductivity

Clicking [**New**], the whole grid will turn blue. Enter the aquitard conductivity values and clicking [**OK**] to accept values.

Now copy the hydraulic conductivity properties of layer 3 to layer 4.

Clicking [**Copy**] from the left toolbar, then [**Layer**].

A dialog box will appear, selecting some options as Figure 4. 28.

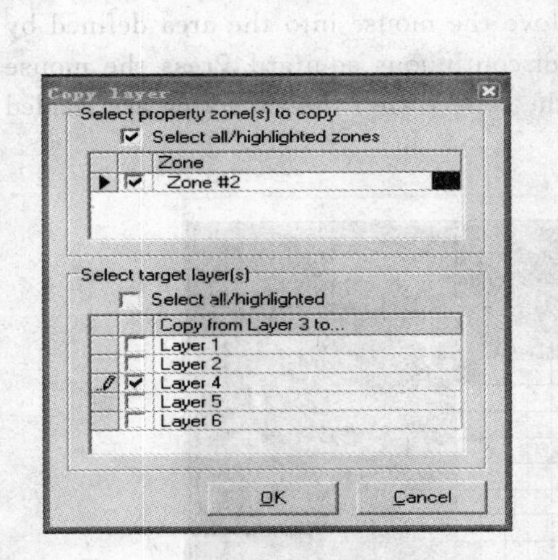

Figure 4. 28 Copy hydraulic conductivity of layer 3 to layer 4 in the model zone

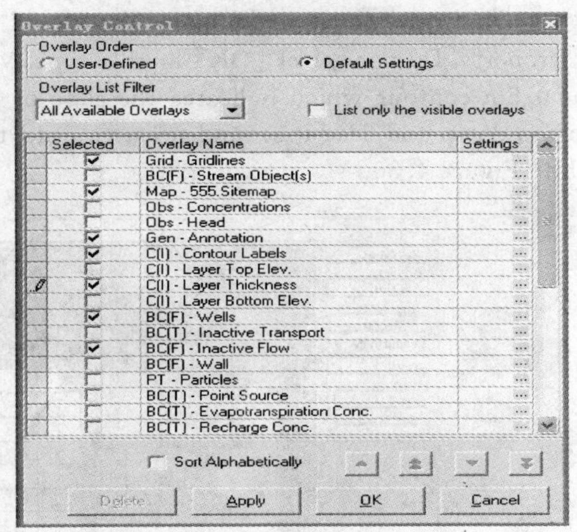

Figure 4. 29 Overlay control

Clicking [**OK**] to copy the K Properties in layer 3 to layer 4.

2. *Display the thickness*

The zone of discontinuous aquitard is delineated on the base map. However, in many cases, this will not be available and you will need to rely on parameters such as layer thickness for determining discontinuous zones. To display the thickness of layer 3 select the [**F9-Overlay**] command button from the bottom toolbar. An **overlay control** dialog box (Figure 4. 29) will appear containing an alphabetized listing of the available overlays which may be displayed when we assign model input parameters.

Scan down the list until you come to **C (I)-Layer Thickness**, Clicking [**OK**] to display a contour plot of the layer thickness.

Now zoom in on the area where the contours converge towards a minimum thickness of 0.5m.

Note! If you are having difficulty viewing the contours because of the background colour follow these steps. You will be able to alter the contour colours from the drop-down list provided.

[**F9-Overlay**]→**C (I)-Layer Thickness**→[...]

Note! To remove the contours of layer thickness, select [**F9-Overlay**] from the bottom menu bar, scan down the list of overlays and deactivate the **C (I)-Layer Thickness** →[**OK**].

3. *Assign K property of the discontinuous aquitard zone*

Clicking [**F5-Zoom In**].

Move the mouse to the upper left of the discontinuous aquitard zone and click the left mouse button. Then stretch a box over the area and click again. A zoomed image of the discontinuous aquitard zone should appear on the screen.

Clicking [**Assign**]→[**Single**] (this assigns a property to an individual cell).

Click on the down arrow under the [**New**] button so that Property #1 is the active *K* property. Do not select [**OK**] at this time. Move the mouse into the area defined by the 0.5m contour which indicates the zone of discontinuous aquitard. Press the mouse button down and drag it around the area until the cells within the zone area are shaded white Figure 4.30.

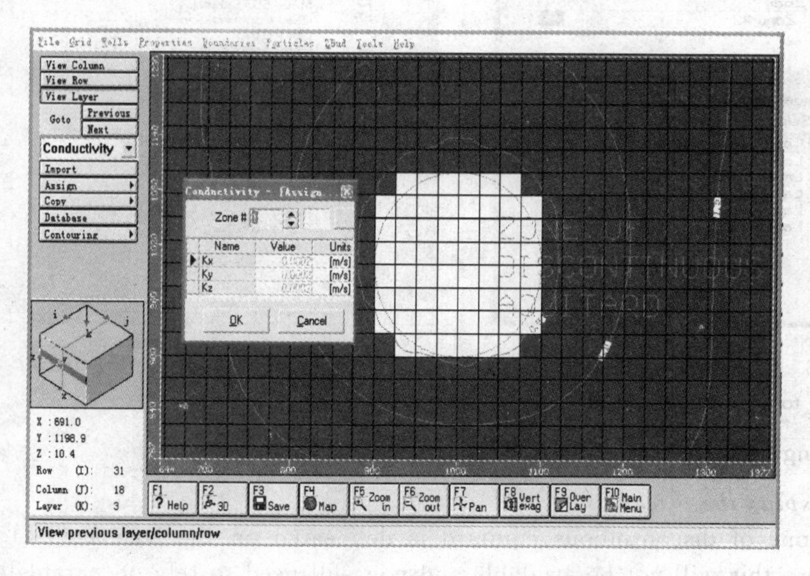

Figure 4.30 Assign *K* property of the discontinuous aquitard zone

Select the [**OK**] button finished "painting" the cells.

Note! If you have re-assigned cells that are outside of the zone, release the left mouse button and press the "right mouse button" to return the cells to their original property color.

Copy this property to layer 4 by Clicking [**Copy**]→[**Layer**].

A dialog box (Figure 4. 31) will appear with the default settings:

Clicking [**OK**].

In order to return to the entire model full-screen display, clicking [**F6-Zoom Out**].

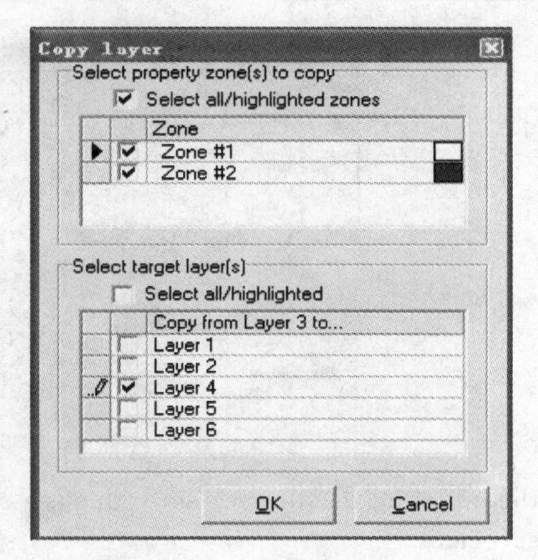

Figure 4. 31 Copy hydraulic conductivity of the discontinuous aquitard zone

4. *To see the model in cross-section*

We select [**View Column**] and then move towards the discontinuous aquitard zone. A cross-section will be displayed as shown in Figure 4. 32.

Return to the plan view display by selecting the [**View Layer**] toolbar button and then clicking on layer 1 in the model cross-section.

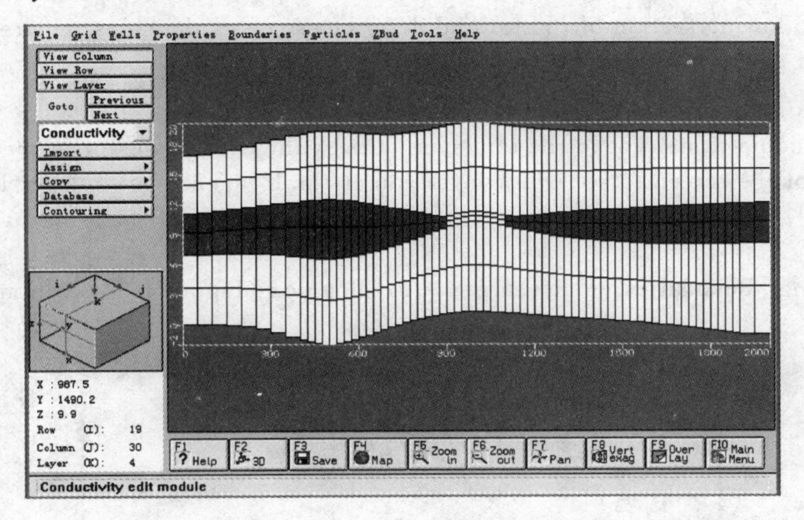

Figure 4. 32 Cross-section shown the discontinuous aquitard zone

5. *Assign storage*

Clicking [Properties]→[Storage]→[Database]. A dialog box (Figure 4. 33) will appear:

Modify the default values and clicking [**OK**].

6. *Initial heads*

The default initial heads condition for a new model can be viewed by pressing [Properties]→[Initial Heads]→[Database].

Visual MODFLOW assume that the initial heads of all layers are constant. However, a better assumption will often significantly decrease the number of iterations required for convergence to occur. For your future models you can use this feature to alter

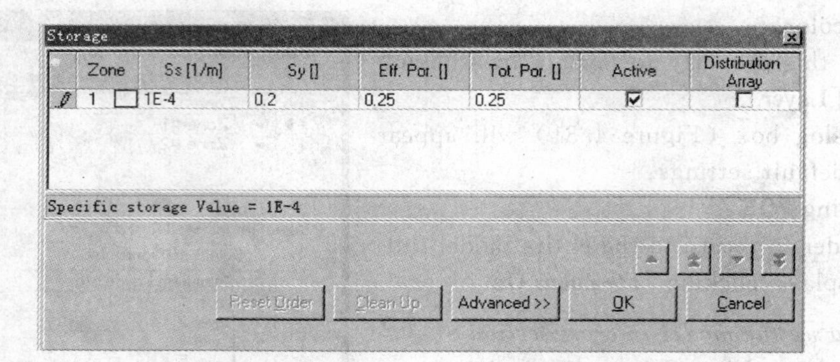

Figure 4. 33 Assigning storage

the initial head estimate; since this is a simple problem, the initial head estimate will be sufficient.

Clicking [**OK**].

4. 3. 5 Assigning Model Boundary Condition

The following section of the tutorial describes some of the steps required to assign the various model boundary conditions.

1. *Assign the recharge condition to the aquifer*

You should firstly get to the top layer. Check the cube navigator in the lower left-hand side of the screen to see which layer you are in. If you are not in layer 1, then use the [Next], [Previous], or [Go to] toolbar buttons in the left-hand menu to advance to layer 1.

Clicking [**Boundaries**]→[**Recharge**]. A dialog box (Figure 4. 34) will appear:

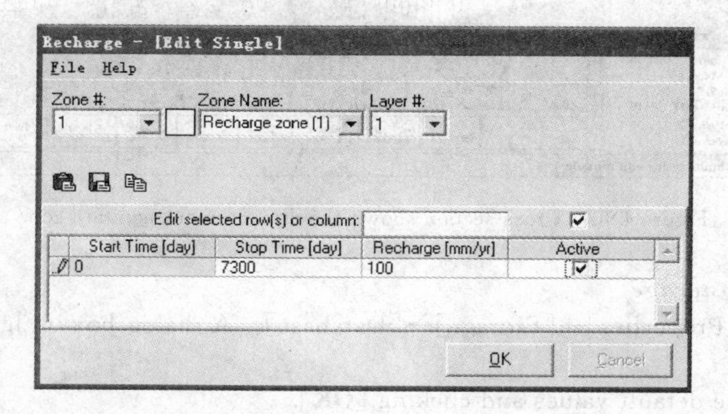

Figure 4. 34 Assigning recharge (edit single)

If you are creating a new model, a Default Recharge dialog box will appear at this point prompting you to enter initial values for recharge. Visual MODFLOW automatically assigns this recharge value to the entire top layer of the model. We enter initial values for recharge and clicking [**OK**] to accept these values.

Then clicking [**F5-Zoom In**], we move the cursor to the upper left of the refueling area and click on the left mouse button. Then stretch a box around the refueling area and click again to close the zoom window. Now assign a higher recharge at the refueling area.

Clicking [**Assign**]→[**Window**].

Move the cursor to a corner of the refueling area. Click the left mouse button, drag a window across the refueling area, and click the left mouse button again. The dialog box will appear, clicking [**New**] and entering some values shown as Figure 4. 35.

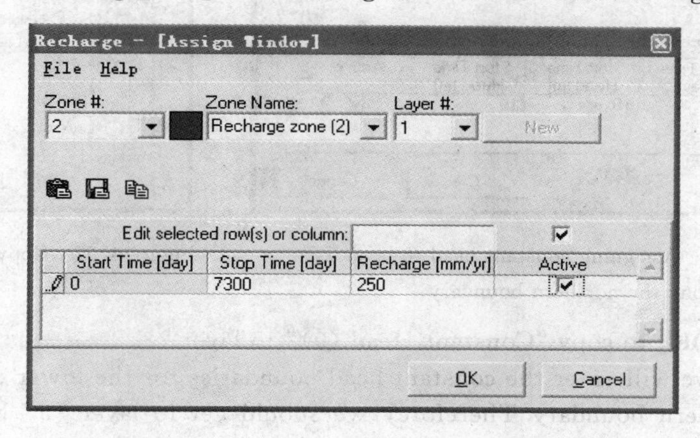

Figure 4. 35 Assigning recharge (whole window)

Clicking [**OK**] to accept these values.

Then we click [**F6-Zoom Out**] in order to return to the full-screen view of the model domain.

2. *Assign constant head boundary conditions*

Clicking [**Boundaries**]→[**Constant Head**], then clicking [**Yes**] to save your Recharge data.

Once your model has been saved, you will be transferred to the constant head Input screen.

Firstly, we will assign constant head boundary condition to the upper unconfined aquifer along the northern boundary of the model domain.

Note! Check the cube navigator in the lower left-hand side of the screen to see which layer you are in (must in layer 1 or layer 2).

Clicking [**Assign**]→[**Line**].

Move the mouse to the north-west corner of the grid. With the left mouse button, we click the north-east corner of the grid with the right mouse button. A horizontal line of cells will be highlighted and an **Constant Head** dialog box will appear as shown in Figure 4. 36.

Enter some necessary values and clicking [**OK**] to accept these values.

The pink line will now turn to a dark red indicating that a constant head boundary value has been assigned.

Secondly, we should copy these conditions to the second layer.

Clicking [**Copy**]→[**Layer**].

The dialog box (Figure 4. 37) will appear, and we select some options as follows:

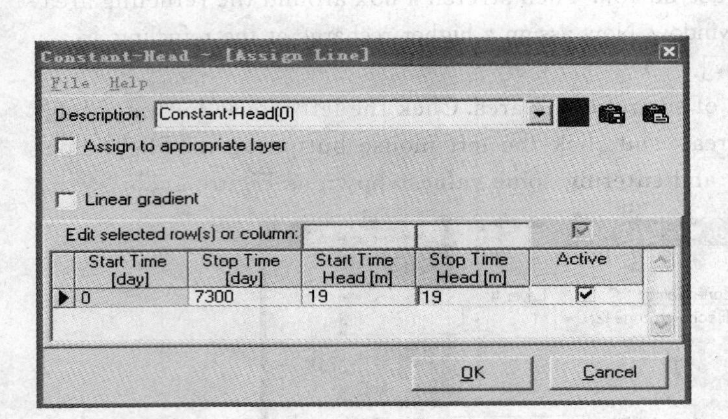

Figure 4. 36 Assigning constant head to layer 1 along the northern boundary

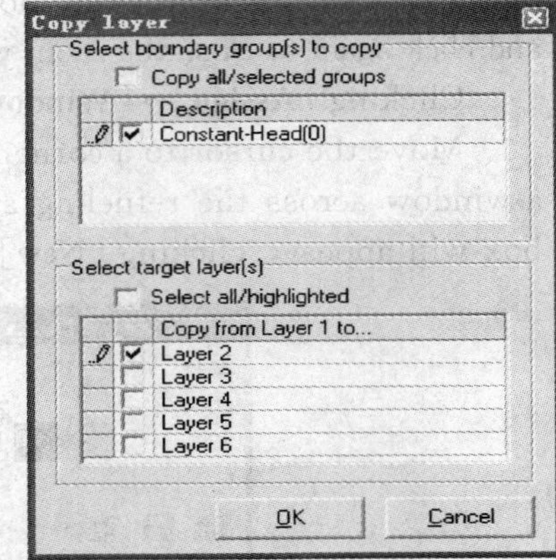

Figure 4. 37 Copy northern boundary of layer 1 and layer 2

Clicking [**OK**] to copy "Constant Head (0)" to layer 2.

Thirdly, we will enter the constant head boundaries for the lower confined aquifer along the northern boundary. Therefore, we should get to layer "5" by clicking [**Go to**].

Clicking [**Assign**]→[**Line**].

Moving the mouse to the north-west corner with the left mouse button, we click the north-east corner with the right mouse button. The line will be highlighted and a dialog box (Figure 4. 38) will appear.

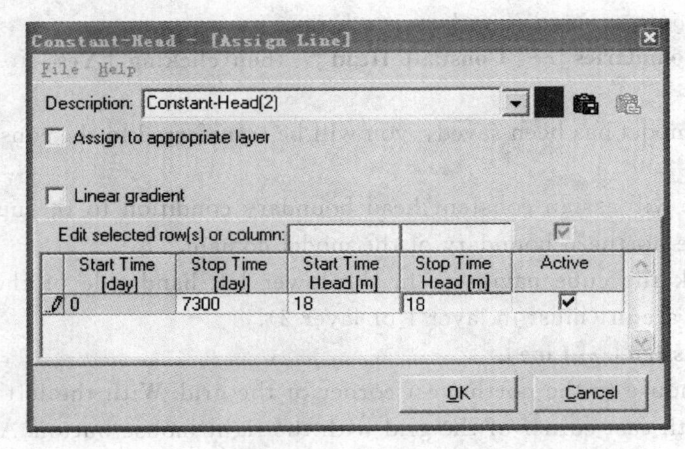

Figure 4. 38 Assigning constant head to layer 5 along the northern boundary

Enter some necessary values and clicking [**OK**] to accept these values.

The pink line will now turn to a dark red indicating that a constant head boundary value has been assigned.

Fourthly, we assign the constant head boundary condition to the lower confined aquifer along the southern boundary of the model domain.

Clicking [**Assign**]→[**Line**].

Moving the mouse to the south-west corner with the left mouse button, we click the south-east corner with the right mouse button. The line will be highlighted and a dialog box (Figure 4. 39) will appear.

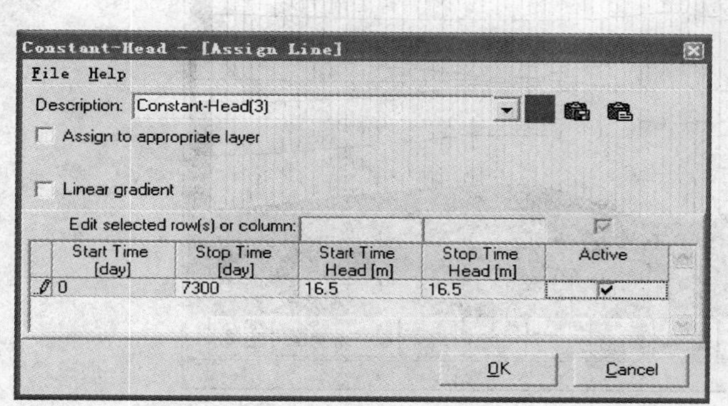

Figure 4. 39 Assigning constant head to layer 5 along the southern bourdary

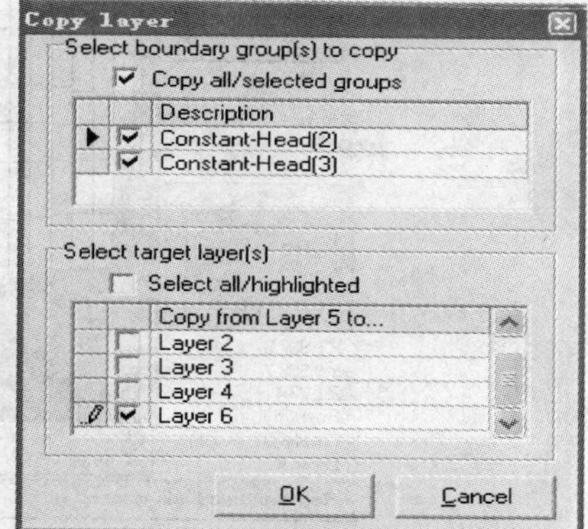

Figure 4. 40 Copy boundary of layer 5 to layer 6

Enter some necessary values and clicking [**OK**] to accept these values.

Fifthly, we should copy these conditions to the sixth layer.

Clicking [**Copy**]→[**Layer**].

The dialog box will appear, and we select some options as shown in Figure 4. 40.

Clicking [**OK**] to copy "Constant Head (2)" and "Constant Head (3)" to layer 6.

After assigning Constant Head values.

Clicking [**View Column**] from the left toolbar.

Move the mouse into the model domain and select the column passing through the zone of discontinuous aquitard. A cross-section similar to Figure 4. 41 should be displayed.

The hydraulic conductivity property colors can also be displayed at the same time. Select [**F9-Overlay**] and a dialog box will appear providing a list of available overlays which may be turned on and off. Using the mouse, double-click on **Prop (F)-Conductivity.** An asterisk will then appear beside the overlay indicating it is now active. Clicking [**OK**] to display the hydraulic conductivity overlay.

3. *Assign a river boundary condition*

Now, we will assign a river boundary condition in the top layer along the southern boundary of the model domain.

Clicking [**Boundaries**]→[**Rivers**] then [**Yes**] to transfer to the rivers input screen.

Clicking [**Assign**]→[**Line**].

Using the sitemap as a guide and along its path, we begin on the south-west side by clicking left mouse button and reached the south-east by click on the right mouse button.

A dialog box [Figure 4. 42 (a) and (b)] will prompt you for information about the river.

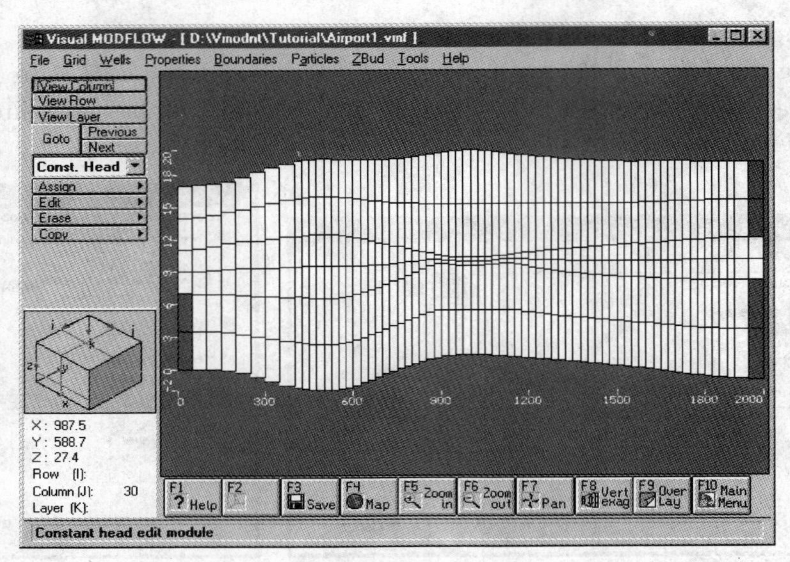

Figure 4. 41 Cross-section of head boundary

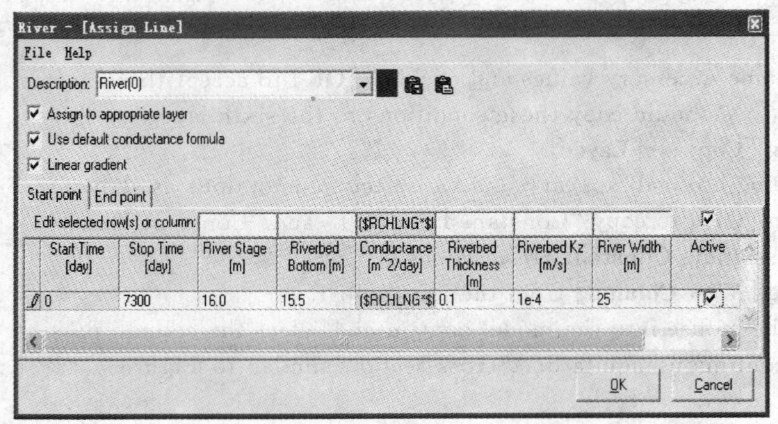

(a) Start point

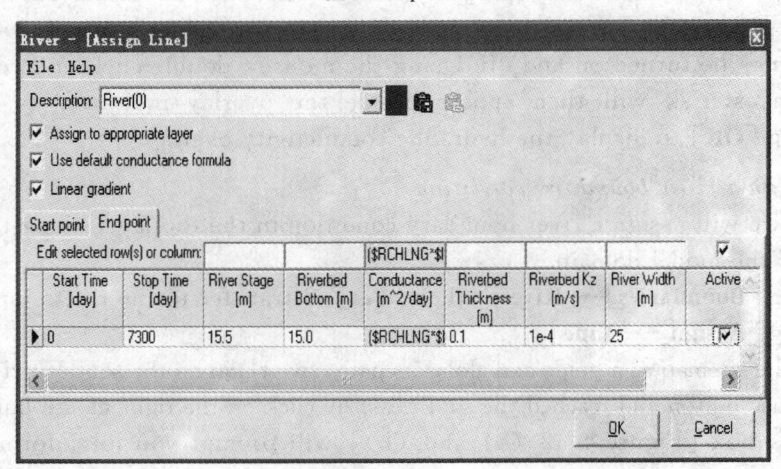

(b) End point

Figure 4. 42 Assigning a river boundary

Clicking [**OK**] to accept these values.

Note! There are two options. "Start point" and "End point" .

After the river has been defined, a blue line will appear delineating the grid cells which have been assigned river boundary conditions as shown in Figure 4. 43.

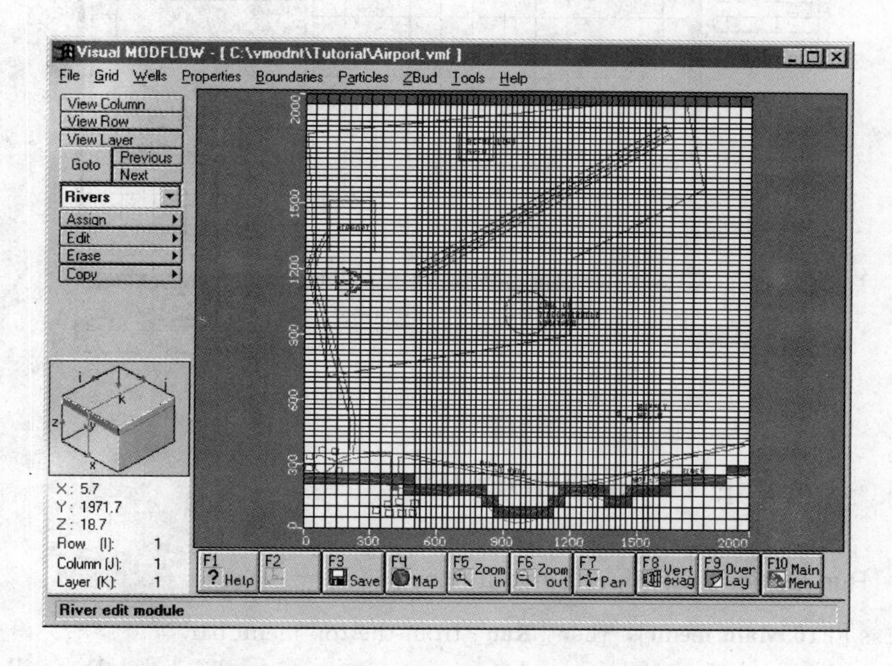

Figure 4. 43 Plane displays the river boundary

4. 3. 6 Assigning Particles

In this section, we will assign forward tracking particles to determine the preferred contaminant exposure pathways.

Clicking [**Particles**] from the top menu bar and will be asked to save your data, press [**Yes**].

The first step should zoom in on the refueling area (the same as previous method).

Clicking [**Add**]→[**Add Line**].

Move the cursor to the left side of the refueling area and click the left mouse button. Stretch a line to the right side of the refueling area then click again. An Add Particles dialog box (Figure 4. 44) will appear. The default number of particles assigned is 10. Change the number of particles to 5.

Clicking [**OK**] to assign the line of 5 particles in the refueling area.

The line of green particles through the refueling area indicates forward tracking particles.

Now return to the full-screen display of the model domain. Press [**F6-Zoom Out**].

Note! The transport model (MT3D Module) have not been mentioned in this book.

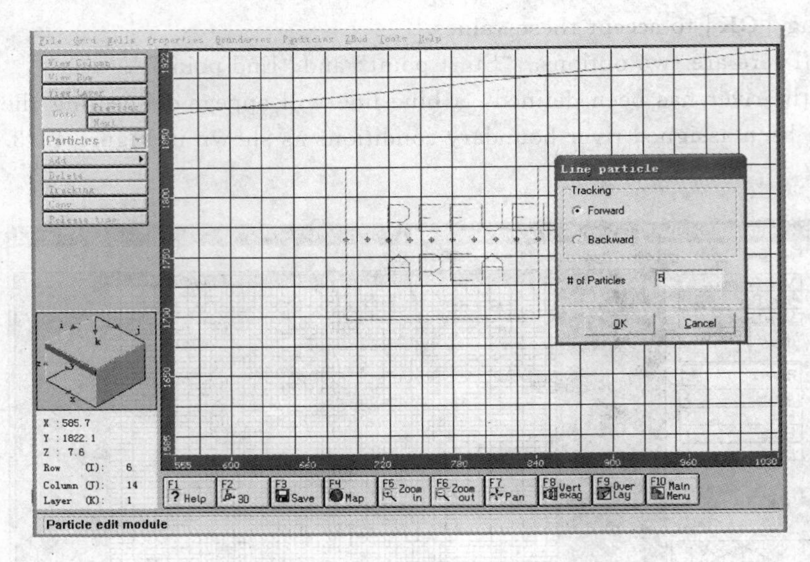

Figure 4.44 Assigning particles

§ 4.4 Running Visual MODFLOW

4.4.1 Run Options for Flow Simulations

Press [**F10 Main menu**], then [**Run**] from the top menu bar.

You will then be transferred to the run options screen and a dialog box will appear as Figure 4.45.

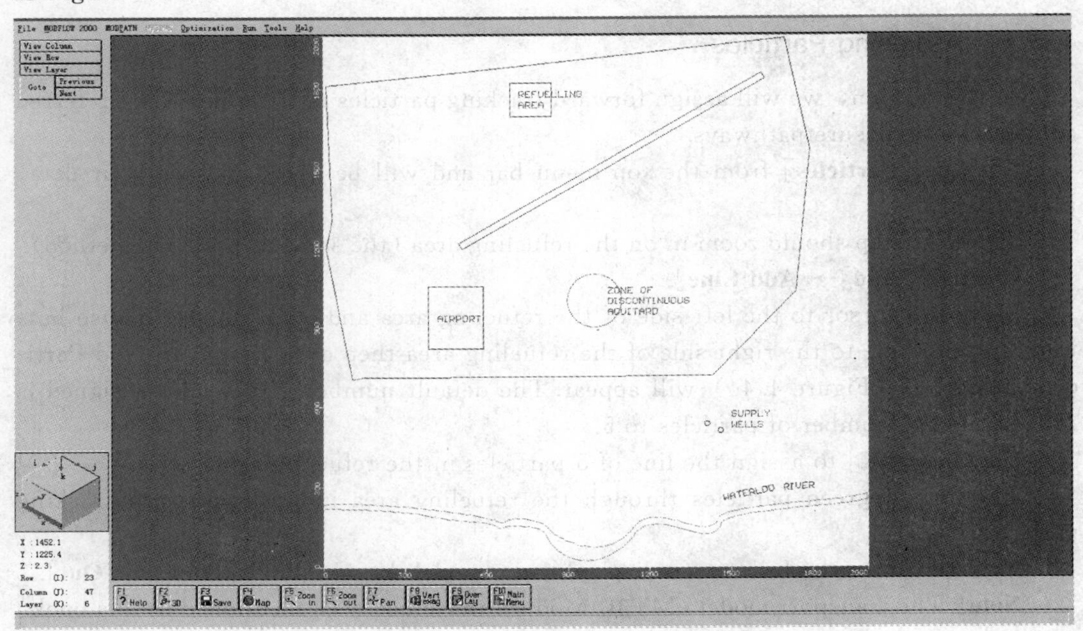

Figure 4.45 Run interface

The run options in Visual MODFLOW are divided into four separate sections; run options for flow simulations (MODFLOW), run options for particle tracking (MODPATH), run options for MT3D simulations (MT3D), and run options for parameter estimation (PEST).

Clicking [**MODFLOW 2000**]→[**Solver**].

The default solver is the WHS Solver, a proprietary solver developed by Waterloo Hydrogeologic Inc. It is the fastest and most stable MODFLOW solver available. For this example, the WHS Solver will be used to calculate the flow solution. When we select the WHS button, A WHS Solver Parameters dialog box (Figure 4. 46) will appear showing all of the default solver settings.

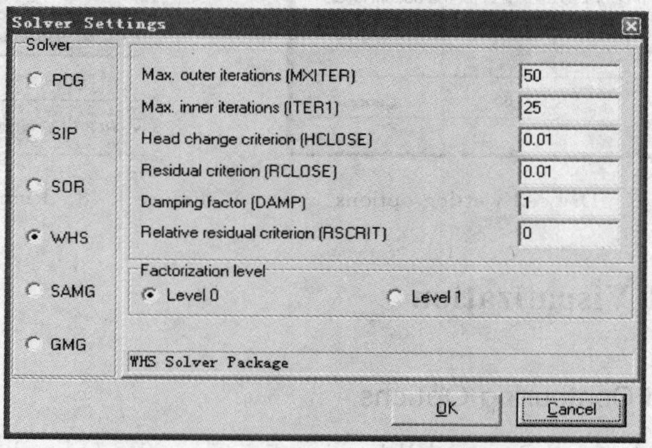

Figure 4. 46 Solver settings

Clicking [**OK**] to accept the WHS Solver settings.

During flow simulations, when water level is lower than the bottom, it is common for cells to run dry and thus become inactive. In order to eliminate this problem from our solution we must activate the cells.

Clicking [**MODFLOW 2000**]→[**Rewetting**], then a **"Dry Cell Wetting Options"** dialog box (Figure 4. 47) will appear:

Select [**Activate cell wetting**] and [**OK**].

There are still several other run options for flow simulations. If you have time, investigate the options which were not mentioned in this tutorial.

4. 4. 2 Engines to Run

If you would like to run the model, Clicking [**Run**] from the top menu bar. A dialog box (Figure 4. 48) will appear.

Select the numerical engines we wish to run as the Figure shown. When we click on the [**Translate & Run**] command button, Visual MODFLOW will create the standard input files for the MODFLOW and MODPATH programs.

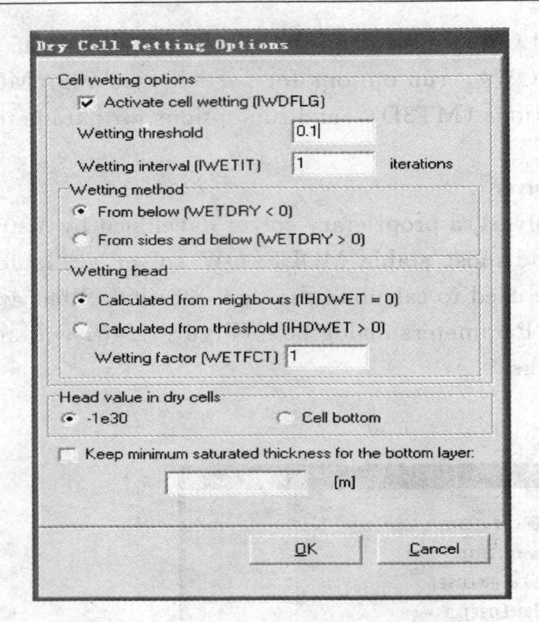

Figure 4. 47 Dry cell wetting options

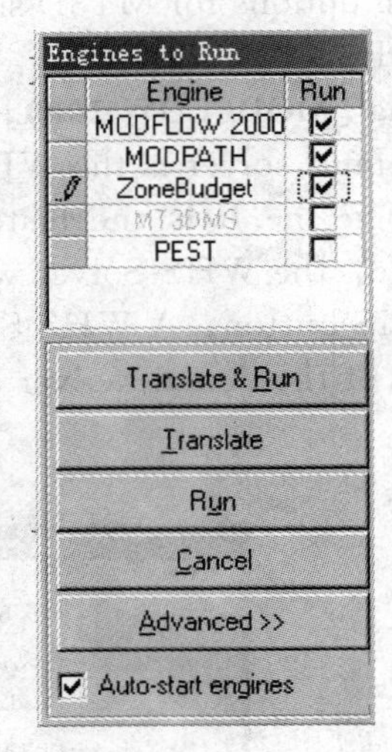

Figure 4. 48 Running engines

§ 4. 5 Output Visualization

4. 5. 1 Head and Contouring Options

After running the model, we will be visualizing results from the model.

Clicking [**Output**], then it is transferred to the Visual MODFLOW Output options screen. By default, the output screen will display a plot of head contours, without color shading, as shown in Figure 4. 49.

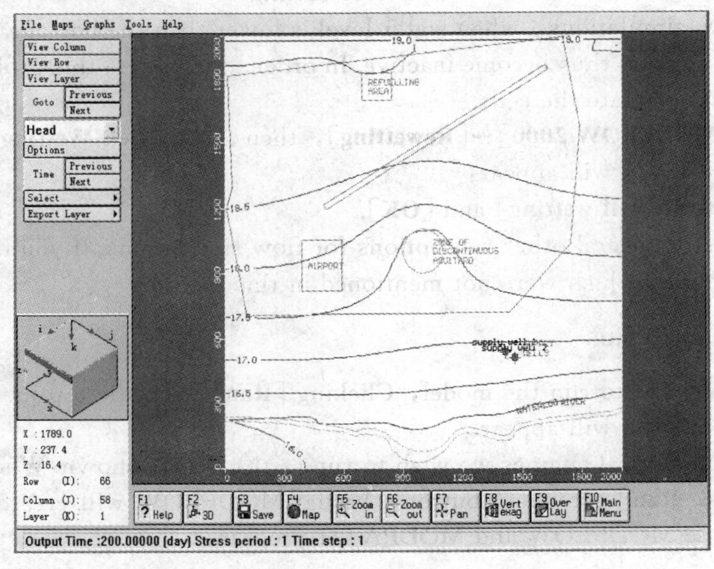

Figure 4. 49 Output interface

To see a list of contouring capabilities, we select [**Option**] from the left menu bar. A dialog box (Figure 4. 50) will appear and enter some values.

You can also change the speed of the contouring by clicking on the button labeled **Contouring resolution/Speed.** You can click on this button once to increase the speed of the contouring by a factor of two. For this exercise, the button should read **High/Slow.**

To activate color shading:

Press [**Color Shading**] tab→[**Use Color Shading**] (Figure 4. 51).

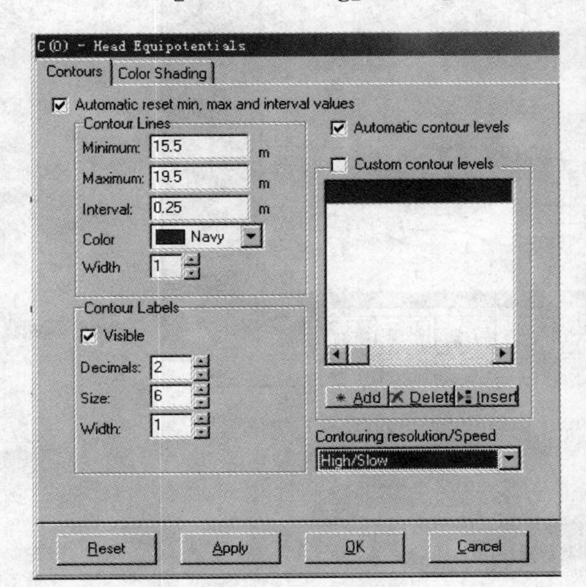

Figure 4. 50 Contours of head equipotentials

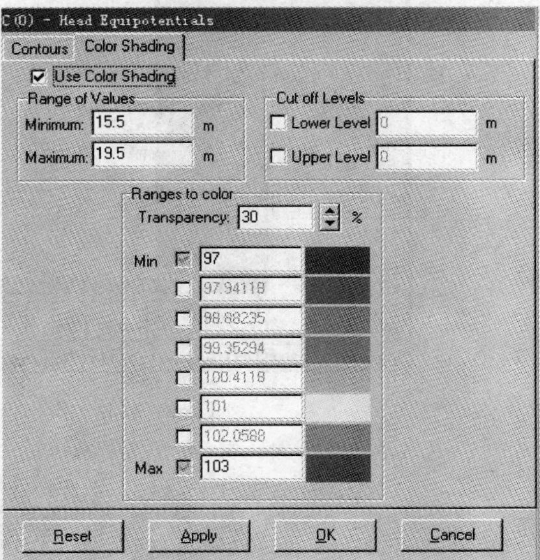

Figure 4. 51 Color shading of head equipotentials

Clicking [**OK**] to accept these contouring options.

The location of the contours should be similar to Figure 4. 52, however, the lines will be slightly more jagged since the contour resolution has been reduced. The Figure 4. 52 does not however show color shading.

Other contouring options that are not activated by the menu buttons can be activated by pressing the **RIGHT MOUSE BUTTON** inside the model domain. This will bring up a shortcut menu (Figure 4. 53) with options to add, delete and move contours and labels.

Select the [**Add Contour**] option and then move the mouse to anywhere in the model domain and click the **LEFT MOUSE BUTTON** once. A contour line will be added in the location that you click the mouse. To add another contour line, simply click the **LEFT MOUSE BUTTON AGAIN.** To finish adding contour lines, click the **RIGHT MOUSE BUTTON** once. To retrieve the shortcut menu, simply press the **RIGHT MOUSE BUTTON** again. This time select the [**Move Label**] option and move the mouse to the location of a contour label, which you wish to move. Press and hold the mouse on the label and then drag the label to the desired location along the contour line and release the mouse button to set the new label location. When you are finished moving contour labels, press the **RIGHT MOUSE BUTTON** once more.

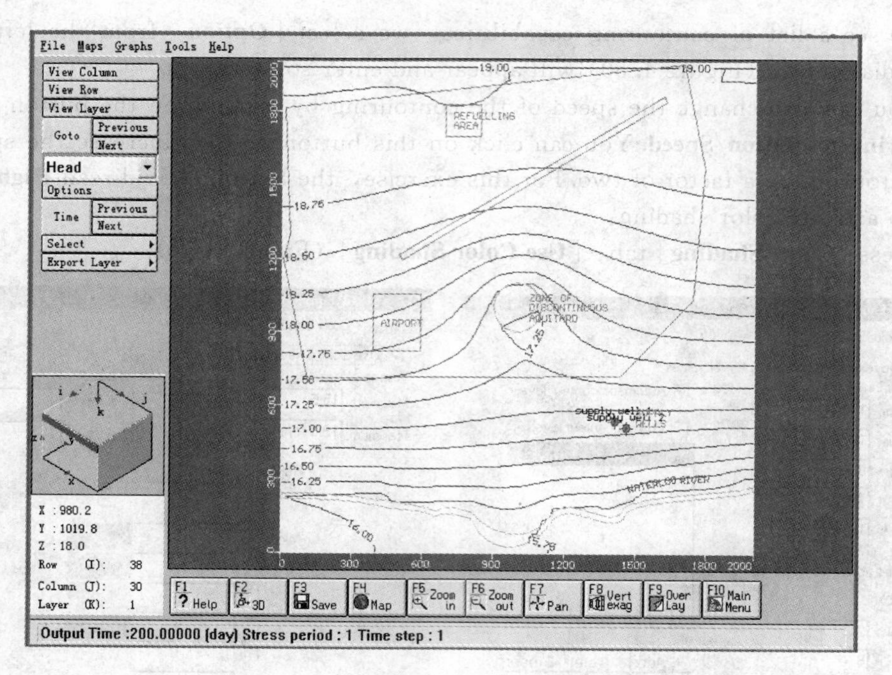

Figure 4. 52 Location of the head contours

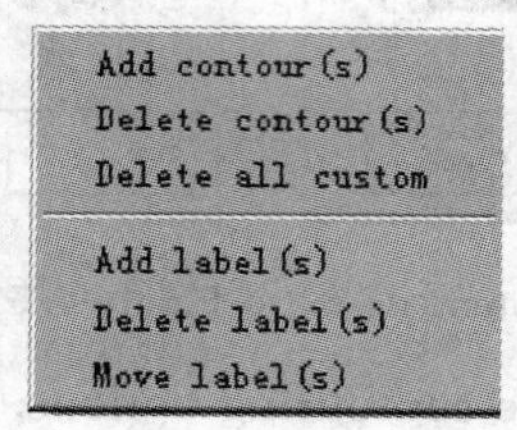

Figure 4. 53 Shortcut menu for the head contours editing

4. 5. 2 Velocity Vectors and Contouring Options

Velocity vectors are an excellent way to visualize the speed and direction of a water particle as it moves through the flow field.

Select [**Velocities**] from the left menu bar.

You will automatically be transferred to the velocity vectors output options screen as shown in the Figure 4. 54.

The velocity vectors will be plotted according to the default settings that plot the vectors in relative sizes according to the magnitude of the flow velocity.

To make the velocity vectors independent of magnitude.

Clicking [**Direction**] from the left toolbar.

All of the arrows will then be plotted as the same size and simply act as flow direction arrows.

Clicking [**Options**] from the left menu. A dialog box will appear and change the number of vectors as shown in Figure 4. 55.

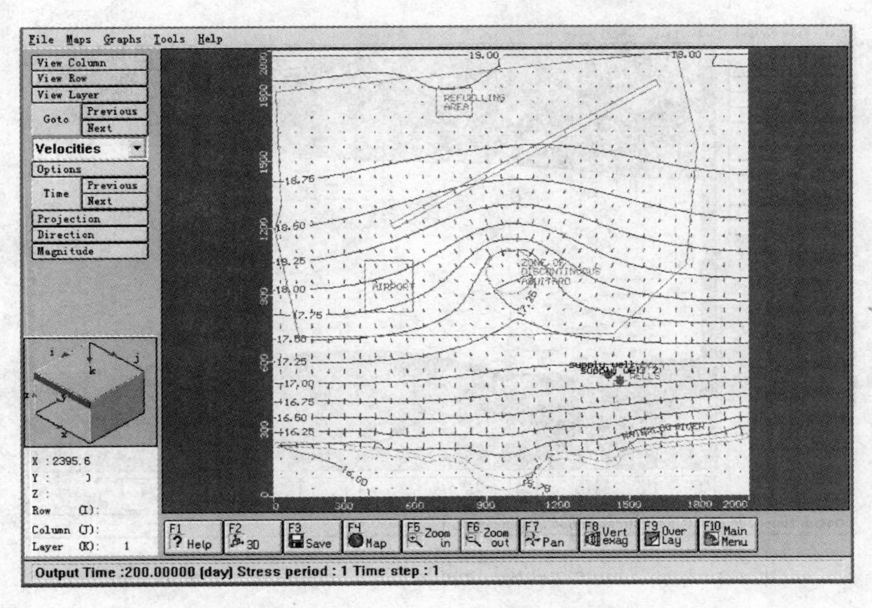

Figure 4. 54 Velocity vectors output options screen

Velocity Options

Color

Vectors : 40 Inward Maroon

Scale factor : 1 Outward Navy

Max. Velocity In Plane Green

Autoscale Scaling Type Fixed scale Variable scale

Reset OK Cancel

Figure 4. 55 Velocity options

Clicking [**OK**] to accept the velocity vector settings.

To see the site in cross-section, select the [**View Column**] button from the left-hand menu and then move the mouse into the model domain and click once to select any column. A display similar to Figure 4. 56 will appear showing both the equipotentials and velocity vectors in cross-section.

To show the discretization of the hydraulic conductivity layers, select the [**F9-Overlay**] button from the bottom menu bar. An Overlay Control dialog box will appear with an alphabetical listing of all the available overlays that you may turn on or off. Locating the **Prop (F)-Conductivity**, double click on it to select it. This should add an asterisk beside the **Prop (F)-Conductivity**, indicating that it has been activated.

Clicking [**OK**] to show the cross-section with the hydraulic conductivity layers (Figure 4. 57).

Return to the plan view display of the model by selecting the [**View Layer**] button

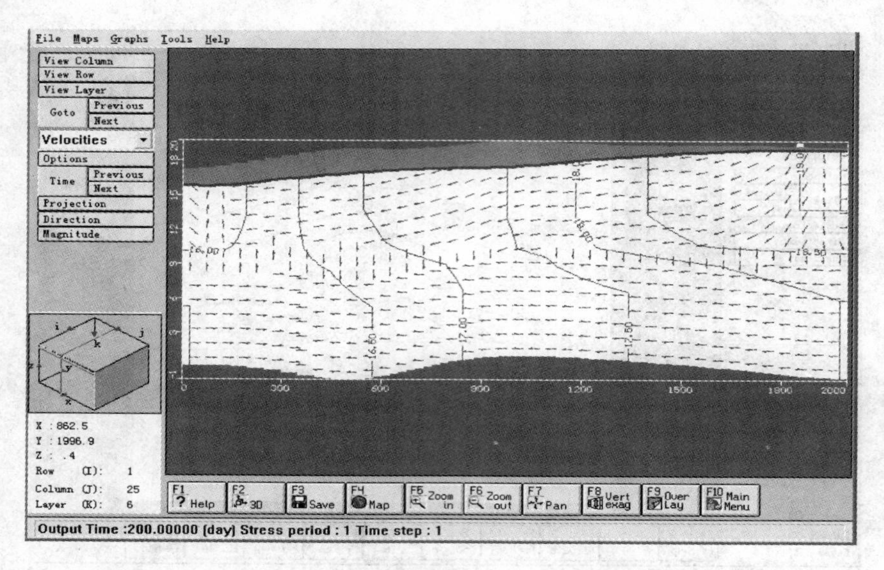

Figure 4. 56 Equipotentials and velocity vectors in cross-section

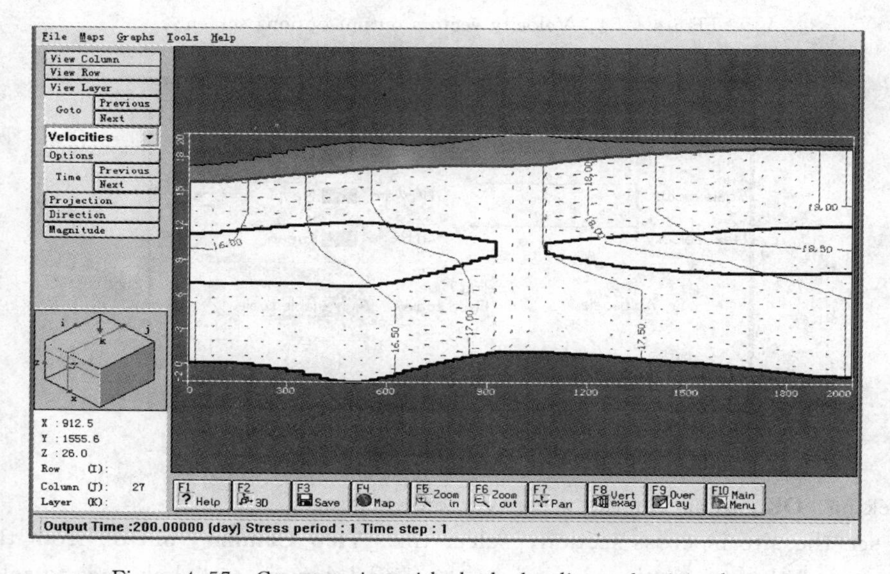

Figure 4. 57 Cross-section with the hydraulic conductivity layers

from the left-hand menu.

To remove the velocity vectors from the screen display select the [**F9-Overlay**] button from the bottom menu bar. An Overlay Control dialog box will appear with an alphabetical listing all of the available overlays that you may turn on or off. Scan down the listing (towards the bottom) and locate the **Vel-Vectors.** Deactivated it.

Clicking [**OK**] to display the screen without the velocity vectors.

4. 5. 3 Pathlines and Pathline Options

Recall that in the input section we assigned five forward tracking particles to the refueling area.

Select [**Pathlines**] from the left menu bar.

The default screen display will plot all of the pathlines as projections from the current layer of the model (Figure 4. 58).

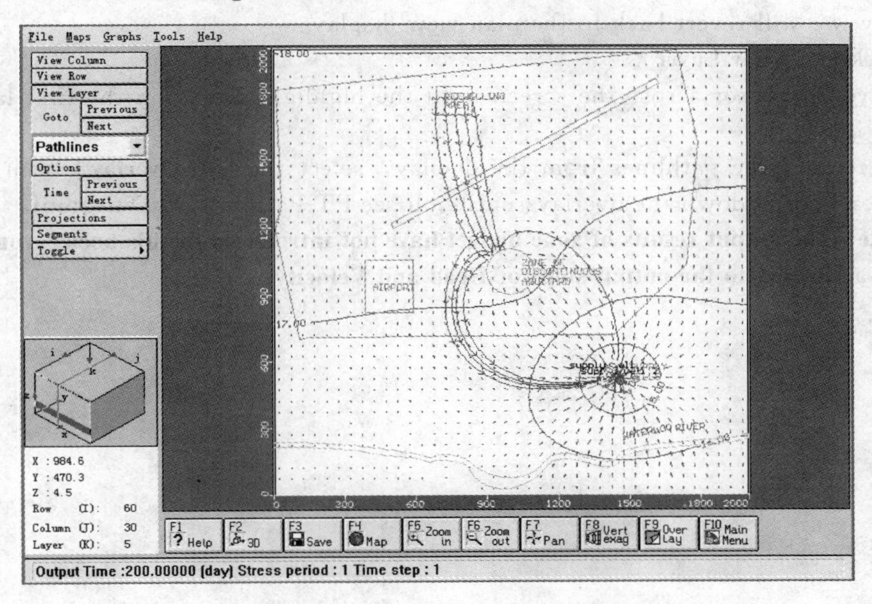

Figure 4. 58 Display of the pathlines

To show the pathlines in the current layer, select the [Segments] button from the left-hand menu. This will show only the pathlines that are in the active layer of the model.

Use the [Next] button to advance through each layer of the model to see where the particle pathlines are located.

Clicking [Projections] to see all pathlines again.

Note! The pathlines have direction arrows on them indicating the flow direction. These arrows also serve as time markers to determine the length of time before a particle reaches a certain destination. To determine the time interval for each time marker, select the [Options] button from the left toolbar and a Pathlines options dialog box will appear as shown in Figure 4. 59.

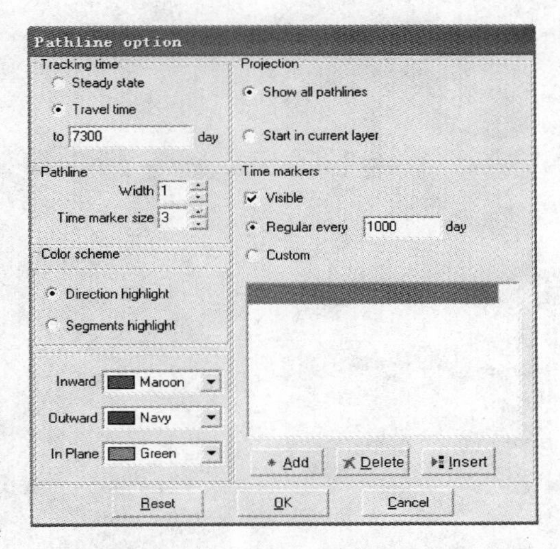

Figure 4. 59 Pathline option

If you want to see how far the pathlines would go in **10000** days, enter a value of 10000 in the box of **Time Related.**

Clicking [**OK**] view the time-related pathlines up to a time of **10000** days.

Viewing the pathlines in cross-section.

Clicking [**View Column**].

Move the cursor across the screen until the highlight bar is in the vicinity of the zone of discontinuous aquitard and click on it.

Now we will revert back to the plan view display.

Clicking [**View Layer**].

Move the cursor across the screen until the highlight bar is on the first layer and click on it.

To remove the pathlines from the display, select the [**F9-Overlay**] from the left toolbar and scroll down the overlays until you see PT-Pathlines. Deactivate it.

Note! The output results of zone budget have not introduced in this book, you can exercise it according to the output steps of Head and Velocity.

Chapter 5 Numerical Simulation of Impervious Wall Construction of Xizhang Basin Groundwater Reservoir in Taiyuan

§ 5. 1 Model Objectives

Groundwater resources play an important role in the national economy, social development and environmental construction of Taiyuan. In the last decade, the development and utilization of groundwater resources have been increased in response to the growing population and rapid development of city construction. At the same time, in the central exploration areas of groundwater (e. g. Xizhang basin), problems appear such as the continual declining of groundwater level, evacuating and pump suspension, ground subsidence, ground water pollution, soil salinization, land desertification, aquifer drainage (water depletion), and a series of environmental and geological problems related to groundwater exploitation.

Xizhang areas, also known as Xizhang basin, are located in northern depressed region of Taiyuan fault basin. There are six main urban water supply sources in the area and the vicinity of the it, which is responsible for 90% water supply of Taiyuan. As far back as 1980, groundwater depression cone has appeared in Xizhang basin. Because of the dramatic decline of the groundwater level, karst water in the upstream of Lancun was withdrew. As a further consequence, springs in Lancun dried up while the karst water level and water output decreased year by year. In order to ensure the sustainable development of the economy, society and environment in Taiyuan, and also to ensure the sustainable utilization of porous groundwater in Xizhang basin, it is necessary to conduct the researches about groundwater resources in Xizhang basin.

Underground reservoir is widely applied in domestic areas as well as abroad. It is viewed as an effective, economical, and environmentally friendly groundwater project. For a long-term advantage, establishing underground reservoir and storing underground water purposefully is an imperative strategy for the sustainable utilization of water resources and the improvement of the environment is in terms of not just water resources but also the whole ecological system in Taiyuan.

§ 5. 2 Overviews of the Xizhang Basin

5. 2. 1 Regional Geography

1. *Location and topography*

Xizhang area is located in the north of Taiyuan, which lies within the longitudes from 112°24′E to 112°35′E and the latitudes from 37°54′N to 38°01′N. It belongs to Jiancaoping district in Taiyuan. The north part of the region is arisen from Lancun and the northern fault zone and the west and east parts touch edge hill. The southern boundary reaches Sangei horst. It is about 13. 5km from east to west and 13km from north to south. The total area is around 109. 54km^2.

The study area is a typical dustpan shaped terrain, which is surrounded by mountains in the east, west and north, but opens to the south. As a result, the study area is a broad alluvial plain. The eastern part of it belongs to Tai-hang Mountains, which is known as Han Mountain or Dong Mountain. The elevation of eastern hill ranges from 1360m to 1700m while the relative height is about from 500m to 800m. The peak of it is called Tian Qiaonao. The northern areas consist of Dong Mountain and Xi Mountain, which peaks at Qizi Mountain with elevation 1418m. The western area, named as Xi Mountain, is a part of Lv Liang Mountain. The topography is precipitous and the elevation ranges from 2000m to 2500m while the relative height varies from 800m to 1900m. There are two intermountain fault basins in the area, i. e. , Huang Da and Ni Tun basin. The fault basin which belongs to Tai Yuan in this area is called Xi Zhang fault basin or northern depressed region, with the elevation from 770m to 820m and the average gradient 1/2000.

2. *Meteorology and hydrology*

Xizhang basin has the features of a continental arid and semi-arid climate. To illustrate, being dry and windy in spring while hot and wet in summer; cool and tend to be waterlogged in autumn whereas cold and with less snow in winter. In summer, the area is mainly affected by the Southeast Asian monsoon, which leads to high precipitation and southeast wind. Conversely, in winter the area is influenced by high pressure of Siberia, which results to less snow and low temperature with northwest wind. The mean annual temperature is 9. 3℃. The average temperature in January and July is －8. 2℃ and 21. 3℃ respectively. The temperature ranges from 32. 5℃ to －21. 7℃. The dusty and windy area is dominated by northwest wind in winter and spring (velocity 2. 5m/s) while by southeast wind in summer and autumn (velocity 1. 9m/s). It is frozen in November and thawed in the March of next year. The freezing depth is ranging from 74cm to 106cm, and the frost-free period is 172 days.

The average annual rainfall is 494mm during 1956 to 2000. Figure 5. 1 describes the precipitation curve of Xizhang basin. The average annual evaporation amount is 3. 6 times that of rainfall, which is about 1806mm. The rainfall is uneven and mainly concentrated in July, August and September, which account for more than 70% of the whole annual amount. The average monthly precipitation is shown in Figure 5. 2.

There are 4 rivers in the Xizhang basin. Fen River is the largest river passing from north to south. Other rivers, such as Yangxing River, Nitun River and Lingjing River, originate from the different region and all flow into Fen River. Due to topography properties and geomorphology conditions in Xizhang basin, karst water supply is the main source of runoff, and the dynamic change of runoff is mainly affected by precipitation.

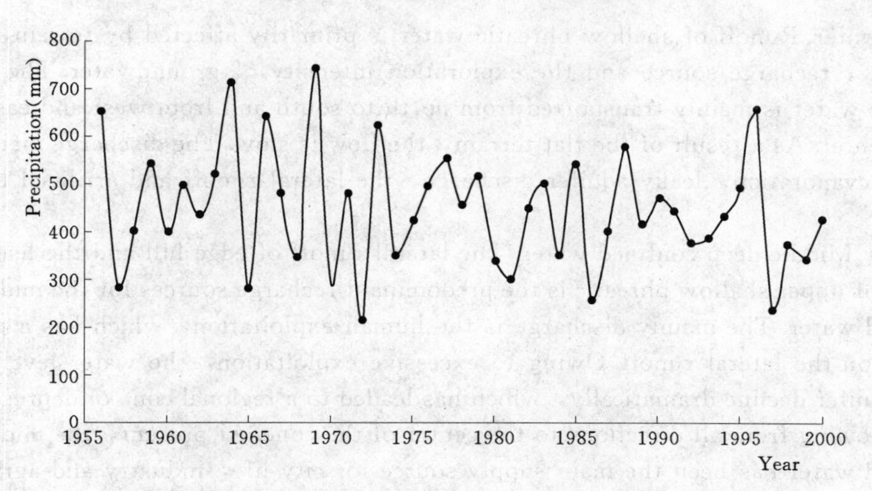

Figure 5. 1 Precipitation curve in Xizhang basin during 1956 to 2000

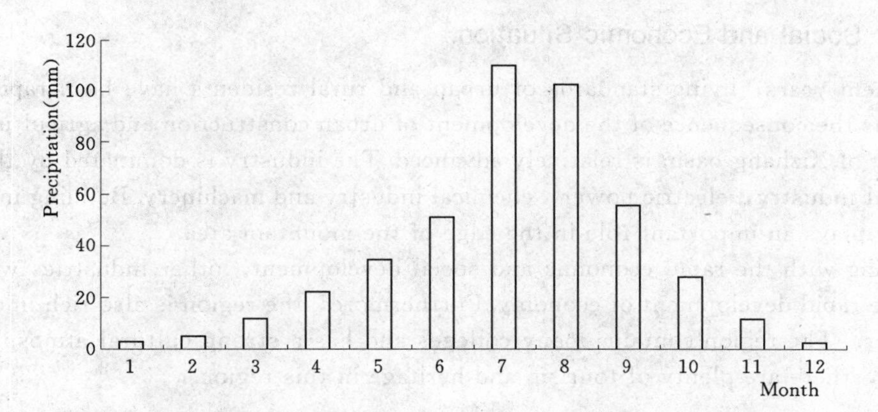

Figure 5. 2 Average month precipitation of several years in Xizhang basin

5. 2. 2 Regional Geological and Hydrogeological Conditions

1. *Regional geology*

The research area is an important part of the Lancun Spring, and the formation of stratum is relatively complete. The stratigraphic distribution is the Paleozoic, the Mesozoic and Cenozoic from the old to the new.

The main water controlling structures are Lancun fault, Tutang fault, Dongjun fault, Boban fault, east arc fault zone and Sangei horst. These water controlling structures have provided the channels through which the surrounding karst water and the fissure water can recharge the pore in Xizhang basin.

2. *Hydrogeological conditions*

The porous water in loosening rocks is divided into shallow phreatic water and middle deep confined water from top to bottom.

(1) Shallow phreatic water. Shallow phreatic water mainly recharged from the atmospheric precipitation, infiltration of canals, irrigation water and lateral runoff of

groundwater. Runoff of shallow phreatic water is primarily affected by terrain, aquifer lithology, recharge source and the exploration intensity of groundwater. The shallow phreatic water is mainly transported from north to south and from west and east to the basin center. As a result of the flat terrain, the flow is slow. The discharge mainly consists of evaporation, leaky aquifer discharge, the lateral runoff and artificial exploitation.

(2) Middle deep confined water. The lateral runoff of edge hill and the leakage recharge of upper shallow phreatic is the predominant recharge sources for the middle deep confined water. The mainly discharge is the human exploitation, which has a great influence on the lateral runoff. Owing to excessive exploitation, the water level of confined aquifer decline dramatically, which has leaded to a regional cone of depression and water flowing from all directions to the center of the cone. At present, the middle deep confined water has been the main supply source for city life, industry and agricultural production.

5.2.3 Social and Economic Situation

Recent years, living standards of urban and rural residents have been rapidly improved as the consequence of the development of urban construction and agriculture. The economy of Xizhang basin is relatively advanced. The industry is dominated by the metallurgical industry, electric power, chemical industry and machinery. Building materials industry plays an important role in the edge of the mountain area.

Along with the rapid economic and social development, other industries will promote the rapid development of economy. Furthermore, the region is also rich in cultural resources. The region contains many colleges and has a strong cultural atmosphere. In addition, there are plenty of tourism and heritage in this region.

§ 5.3 Conditions for the Construction of Underground Reservoir

1. *Suitable hydrogeological conditions*

Underground reservoir in Xizhang basin lies in the basin controlled by the structure around and stratigraphic lithology. The three sides are recharge boundary and the south is Sangei horst. The south of the shallow phreatic groundwater is the discharge boundary; the middle deep confined water is affected by geological structures and in some sections the water level lines are perpendicular to main tectonic lines, and the boundary is flow lines (impermeable boundary). The underground reservoir water storage space is the pore in the Quaternary loose deposit, the thickness of water storage formation is large and stable, and the groundwater storage is large. The aquifer lithologies are mainly of sand and gravel, which have considerable water supply and regulation capacity.

2. *Water supply resources*

Fen River passes from north to south through the basin. In the irrigation periods of winter and spring, the upstream discharge and of Fen River Reservoir is the main supply source for the underground reservoir. Other tributaries of Fen River in the region have a large watershed area, which mainly recharges the groundwater when flood happens in the area. In addition, precipitation and treated sewage infiltration is also available water supplies for underground reservoir.

3. *Artificial recharge measures for groundwater*

There is no thick regional aquifer existing in vadose zone and the surface infiltration capacity in Xizhang basin is large. The deposition favorable positions (such as Ancient River, modern riverbed) contribute to better vertical infiltration for surface water and precipitation. Concerning the stratigraphic lithology characteristics and geological structure, two supply measures are applied — Firstly, dredging rivers and channels before the flood season on Fen River, Yangxing River and Nitun River to increase the recharge capacity for groundwater; Secondly, constructing an impervious wall in Snagei horst to intercept groundwater runoff and increase the available groundwater resources.

4. *Exploration measures*

Xizhang basin is the main exploration district for Taiyuan urban water supply. However, the excessive exploitation results in the regional declining of water level and formed a larger area of depression cone. For a longer period in the future, groundwater will still be the main water supply source for Taiyuan. There have been 306 mining wells existing in the area, so the region is capable of exploration.

5. *Benefits for social well-being, environment and economy*

The construction of underground reservoir should satisfy the local production, domestic water demands and promote development of social economy. During the construction and operation of underground reservoir, we should consider not only increasing the investment of per unit of water project, but also the water cost, the comprehensive income of aspects and a variety of favorable and unfavorable impacts on the environment.

§ 5.4 Groundwater Flow Model for Xizhang Basin

5.4.1 Hydrogeological Conceptual Model

According to the distribution of depression cone, the simulation area was chosen based on the water level contour map of shallow and middle deep groundwater level in 2005. As shown in Figure 5.3, the area is about 13.5km from east to west and 13km from north to south with the total area around 109.54km^2.

Based on the conditions of hydrogeological and groundwater exploitation, the aquifers are divided into three layers from top to bottom, that is shallow phreatic aquifer, an aquitard and middle deep confined aquifer. Since phreatic water level is higher than the confined aquifer, the leakage recharge from shallow phreatic will be transported through the aquitard to middle deep confined aquifer. Affected by natural conditions and artificial exploitation, water level is unsteady in the simulation area. Aquifer is quaternary porous medium and can be generalized as heterogeneous and isotropic media. On the basis of systematic analysis on the basic conditions and drawing on previous work of parameters partition, the same parameters partition can be regarded as homogeneous layer and the flow obey Darcy's law. According to the geology and hydrogeology, and hydrodynamic characteristics of shallow and middle deep groundwater, the boundaries of the simulation area are generalized as the second type flow boundary. Based on the actual long-term observed water level, simulation period is from January 2005 to December 2005, and the water head distribution on January 1, 2005 is selected as the initial flow field.

There are two common methods to determine the hydrogeological parameters. First, it can be obtained from the field test. Second, it is worked out through long-term groundwater observation data. The initial input parameters mainly come from the results of this test and the previous researches. We use Visual MODFLOW software to undertake the new parameter calibration and partition.

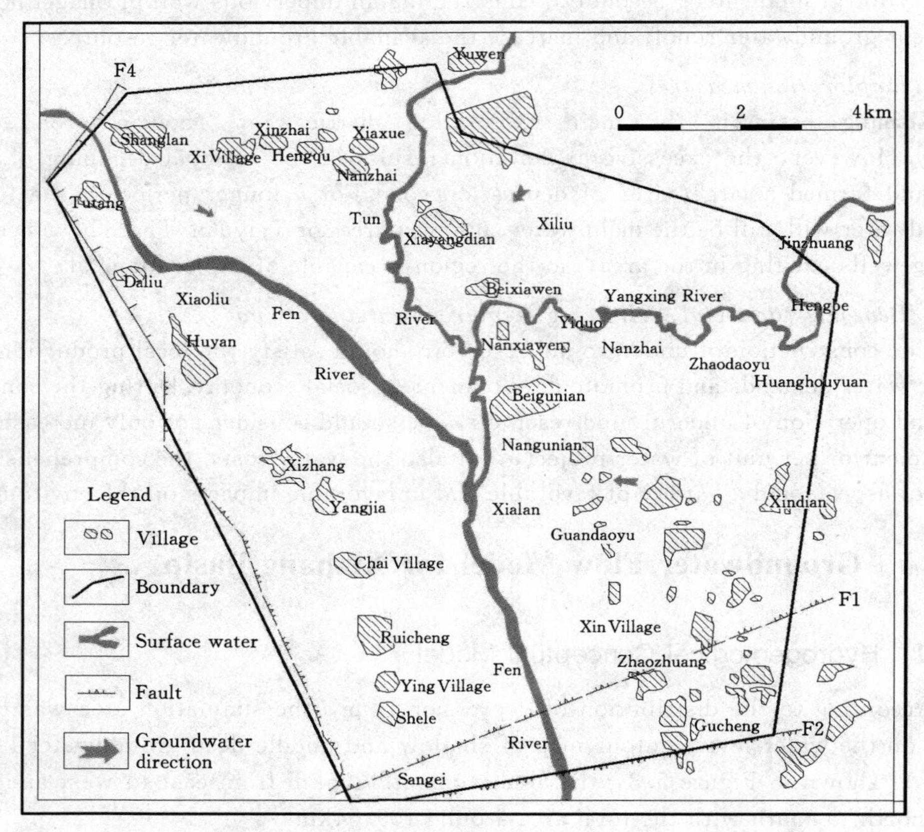

Figure 5.3 Research area

5.4.2 Establishment Groundwater Model

1. *Establishment of mathematical model for groundwater flow*

Based on the hydrogeological conceptual model of groundwater systems, the mathematical models for water level distribution in the simulated area are established. Ignoring the storage capacity of aquitard, the mathematical model of shallow phreatic water and middle deep confined water are defined as:

Model of shallow phreatic water

$$\begin{cases} \frac{\partial}{\partial x}\left[K(H_1-B)\frac{\partial H_1}{\partial x}\right]+\frac{\partial}{\partial y}\left[K(H_1-B)\frac{\partial H_1}{\partial y}\right]+\frac{\partial}{\partial z}\left[K(H_1-B)\frac{\partial H_1}{\partial z}\right] & \\ \quad +\frac{K'}{m'}(H_1-H_2)+q_1=\mu\frac{\partial H_1}{\partial t} & (x,y,z)\in(D),t\geqslant 0 \\ h(x,y,z,t)|_{t=0}=h_0(x,y,z) & (x,y,z)\in(D) \\ K(h-b_1)\frac{\partial h}{\partial \overline{n}}=q_1(x,y,z,t) & (x,y,z)\in(\Gamma_1),t>0 \end{cases} \tag{5.1}$$

Model of middle deep confined water

$$\begin{cases} \frac{\partial}{\partial x}\left[T\frac{\partial H_2}{\partial x}\right]+\frac{\partial}{\partial y}\left[T\frac{\partial H_2}{\partial y}\right]+\frac{\partial}{\partial z}\left[T\frac{\partial H_3}{\partial z}\right] & \\ \quad +\frac{K'}{m'}(H_2-H_1)+q_2=\mu_s\frac{\partial H_2}{\partial t} & (x,y,z)\in(D),t\geqslant 0 \\ h(x,y,z,t)|_{t=0}=h_0(x,y,z) & (x,y,z)\in(D) \\ T\frac{\partial h}{\partial \overline{n}}=q_1(x,y,z,t) & (x,y,z)\in(\Gamma_2),t>0 \end{cases} \tag{5.2}$$

Where: K is the average permeability coefficient of shallow groundwater, $[LT^{-1}]$; T is the hydraulic conductivity of mid-depth groundwater subsystem, $[L^2T^{-1}]$; H_1 shows the water level of shallow phreatic water, $[L]$; H_2 shows the water level of middle deep groundwater, $[L]$; μ indicates the specific yield of phreatic water; μ_s stands for the storage coefficient of confined aquifer; m' defined as the thickness of aquitard, $[L]$; K' defined as the permeability coefficient of aquitard, $[LT^{-1}]$; q_1 is the algebraic sum of vertical recharge intensity for shallow groundwater, $[LT^{-1}]$; q_2 is the pumping intensity of middle deep groundwater, $[LT^{-1}]$; B represents the top boundary elevation of aquitard, $[L]$; D represents the plane seepage zone which is determined by shallow and middle deep groundwater simultaneously, $[L^2]$; Γ_1 and Γ_2 are the flow boundary of, $[D]$; $\overline{n}$ is the outer normal vector of second type boundary.

And we solve the models with backward difference method and iterative solution.

2. *Subdivision of simulation area*

After establishing the groundwater flow model, we will adopt Visual MODFLOW software to solve the groundwater flow.

As shown in Figure 5.4, the simulation area is subdivided into 130 rows and 138 columns, and the number of the rectangle cell is 53820. The active cells are 32913 while the inactive cells are 20907. The area of each cell is 0.01km^2. According to Figure 5.4, hydrogeological units are classified into 3 layers in the vertical direction (Figure 5.5). The active cells are represented with white while the inactive cells are represented with gray.

3. *Boundary conditions*

Initial condition: According to the monitoring data of long-term wells on January 1, 2005, the flow field of phreatic water and middle deep confined water is obtained by

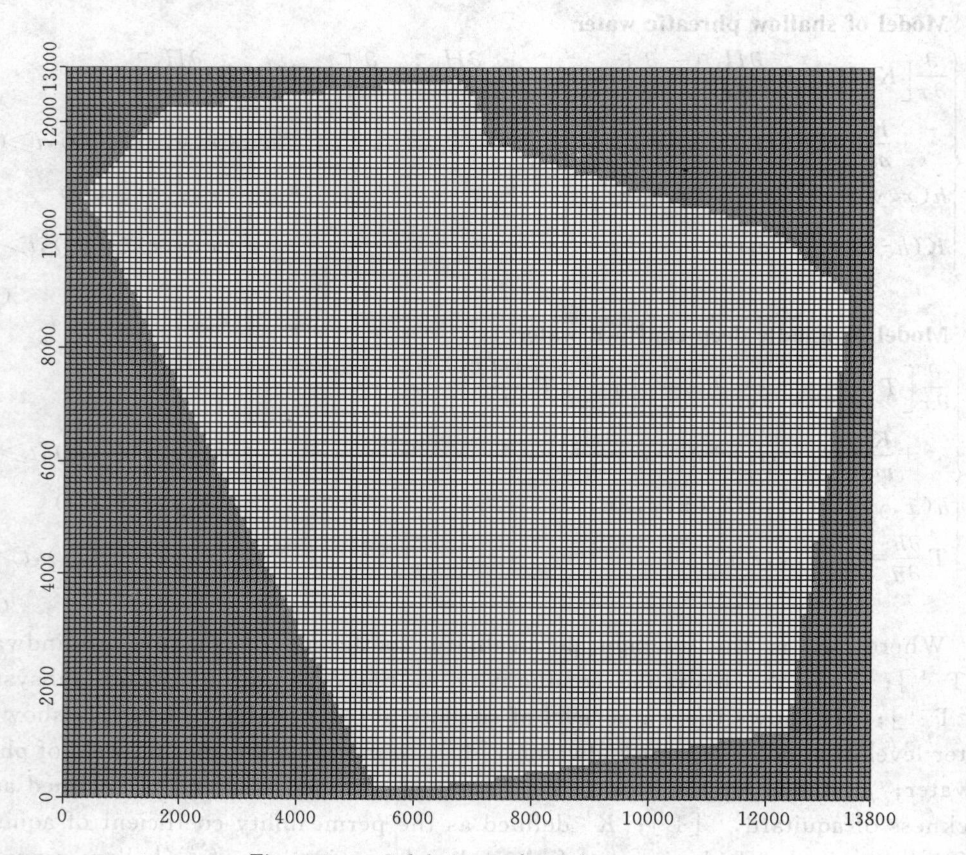

Figure 5. 4 Subdivision of simulation area

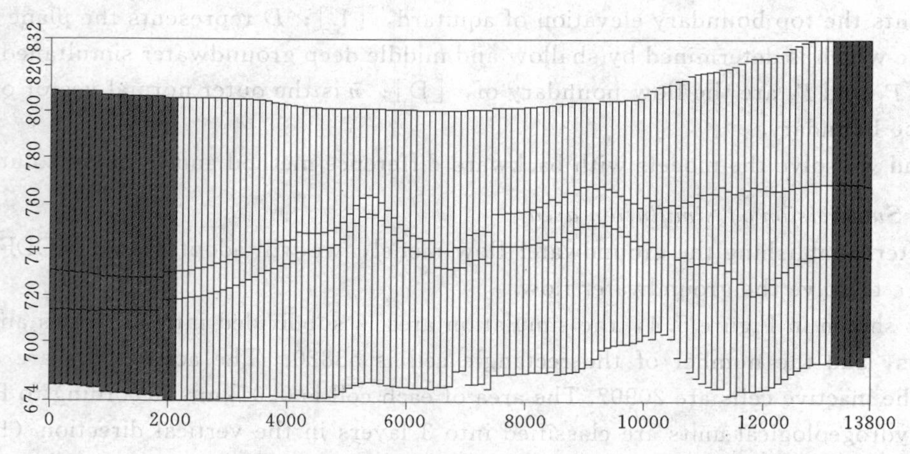

Figure 5. 5 Finite-difference discretion in vertical

interpolation calculation, which is selected as the initial flow field. The results are shown in Figure 5. 6.

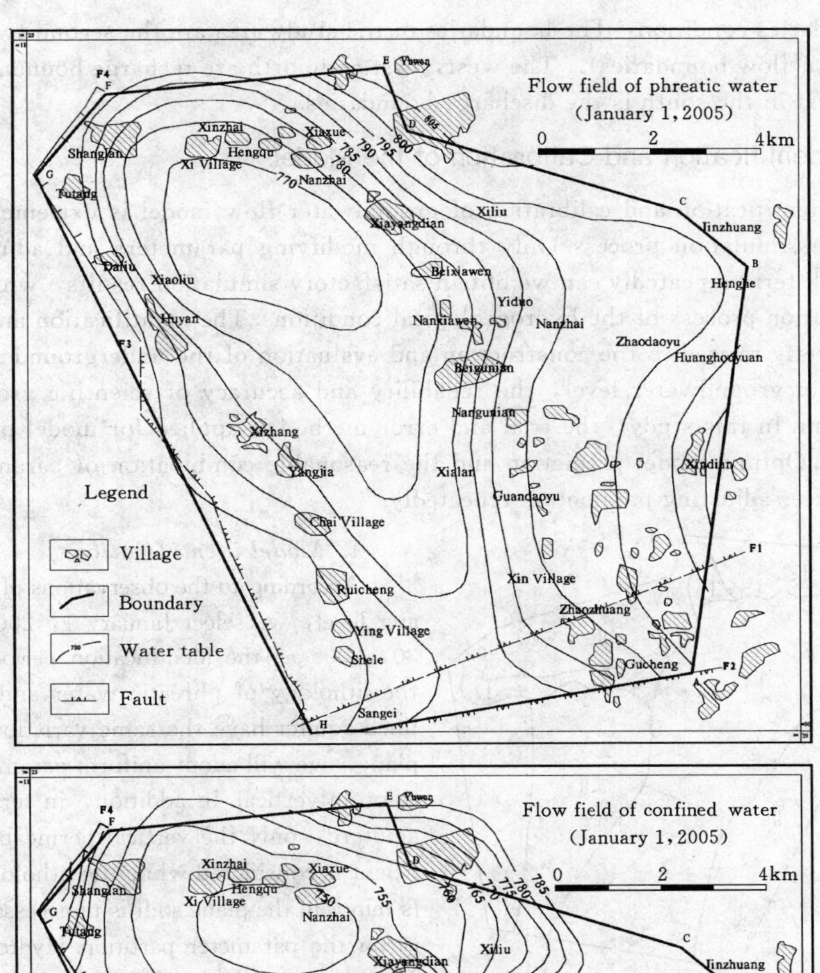

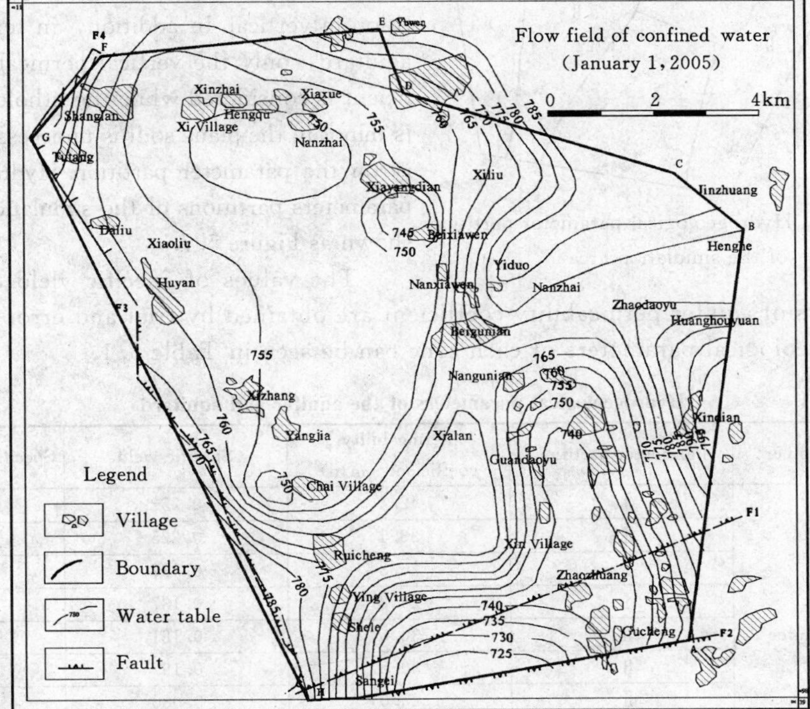

Figure 5.6 Initial groundwater flow field of Xizhang basin (on Jan. 1, 2005)

Boundaries condition: The boundaries of the study area are the second type boundaries (i. e. , flow boundaries). The west, east and north are recharge boundaries while Sangei horst in the south is the discharge boundaries.

5.4.3 Identification and Calibration of the Model

The identification and calibration of groundwater flow model is extremely significant in the simulation process. Only through modifying parameters and adjusting the source/sink term repeatedly can we obtain satisfactory simulation results, which is also the recognition process of the hydrogeological conditions. The identification and calibration is directly related to the construction and evaluation of the underground reservoir, prediction of groundwater level, the reliability and accuracy of scientific groundwater management. In this study, the trial and error method is applied for model parameters calibration. Optima model prediction and the reasonable combination of parameters are obtained after adjusting parameters repeatedly.

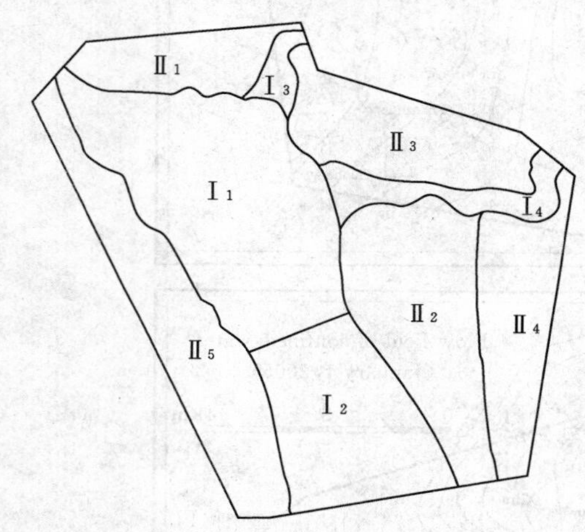

Figure 5.7 Hydrogeological parameter partition of the simulation area

1. *Model identification*

According to the observations of groundwater level, we select January 1, 2005 to June 30, 2005 as the identification period. Because the lithology of phreatic water and the confined aquifer have the same variation trend in plane, we will adopt uniform parameters partition in vertical. In addition, in terms of the aquitard, only the vertical permeability coefficient is considered while the lithology change is minor in the plane so it is unnecessary to carry on the parameter partition. Hydrogeological parameters partitions of the simulation area are shown as Figure 5.7.

The values of specific yield and partitions results of aquifer permeability coefficient are obtained by trial and error methods, and hydrogeological parameters of each zone can be seen in Table 5.1.

Table 5.1 Hydrogeological parameters of the aquifer and aquitard

Stratum number	Parameter partition	Permeability coefficient (m/d)	Specific yield	Specific storativity
Phreatic aquifer	Ⅰ1	42.4	0.253	
	Ⅰ2	33.1	0.225	
	Ⅰ3	26.3	0.137	
	Ⅰ4	26.3	0.137	
	Ⅱ1	30.5	0.181	
	Ⅱ2	30.5	0.197	
	Ⅱ3	25.0	0.080	
	Ⅱ4	25.0	0.110	
	Ⅱ5	6.6	0.078	

Continued

Stratum number	Parameter partition	Permeability coefficient (m/d)	Specific yield	Specific storativity
Aquitard		0.15		0.000020
Confined aquifer	I_1	62.5		0.000020
	I_2	46.3		0.000013
	I_3	36.7		0.000017
	I_4	36.7		0.000011
	II_1	54.6		0.000016
	II_2	54.6		0.000016
	II_3	22.3		0.000011
	II_4	22.3		0.000007
	II_5	11.2		0.000009

The water head in the observation well is calculated by the mathematical model. Through comparing with the observed water head, the hydrogeological parameters can be computed reversely. We select 8 representative water level observations to conduct the simulation, which includes 5 phreatic aquifer and 3 confined aquifer. The water level fitting curves between the observed head and simulated values are shown as Figure 5.8.

The errors (difference between the observed head and is simulated values) that are less than 0.3m account for 70% of the whole errors, which is consistent with the identification requirements. The simulation results show that the positive-negative errors are more even and the system is stable. Generally speaking, the accuracy and effectiveness of the simulation is ideal.

2. *Model validation*

To validate the reliability of hydrogeological parameter and the model, the calibrated model and parameters are applied to simulate the groundwater level of every observed well and then the results are compared with the observed heads. According to actual observed data, validation period is set from July 1, 2005 to December 31, 2005. The comparisons between the observed head and calculated values are shown as Figure 5.9.

The errors less than 0.28m account for 72% of the total errors, which are consistent with the validation requirements. The results show that through conceptualizing the geological conditions, determining the boundary conditions and selecting the hydrological parameters values. The model can appropriately depict the hydrogeological characteristics and be applied for forecasting of the groundwater level and the management of groundwater resources.

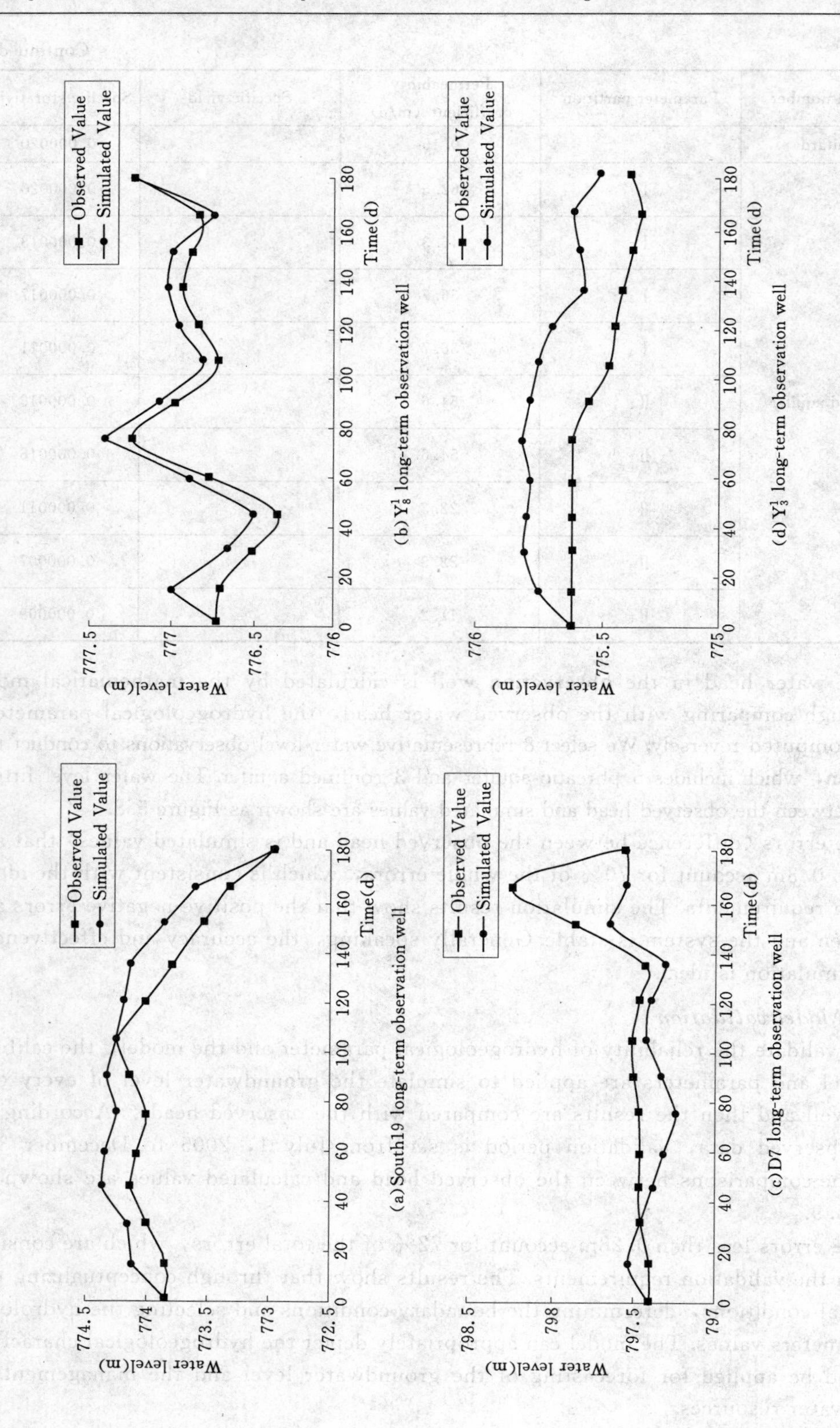

(a) South19 long-term observation well

(b) Y_8^1 long-term observation well

(c) D7 long-term observation well

(d) Y_3^1 long-term observation well

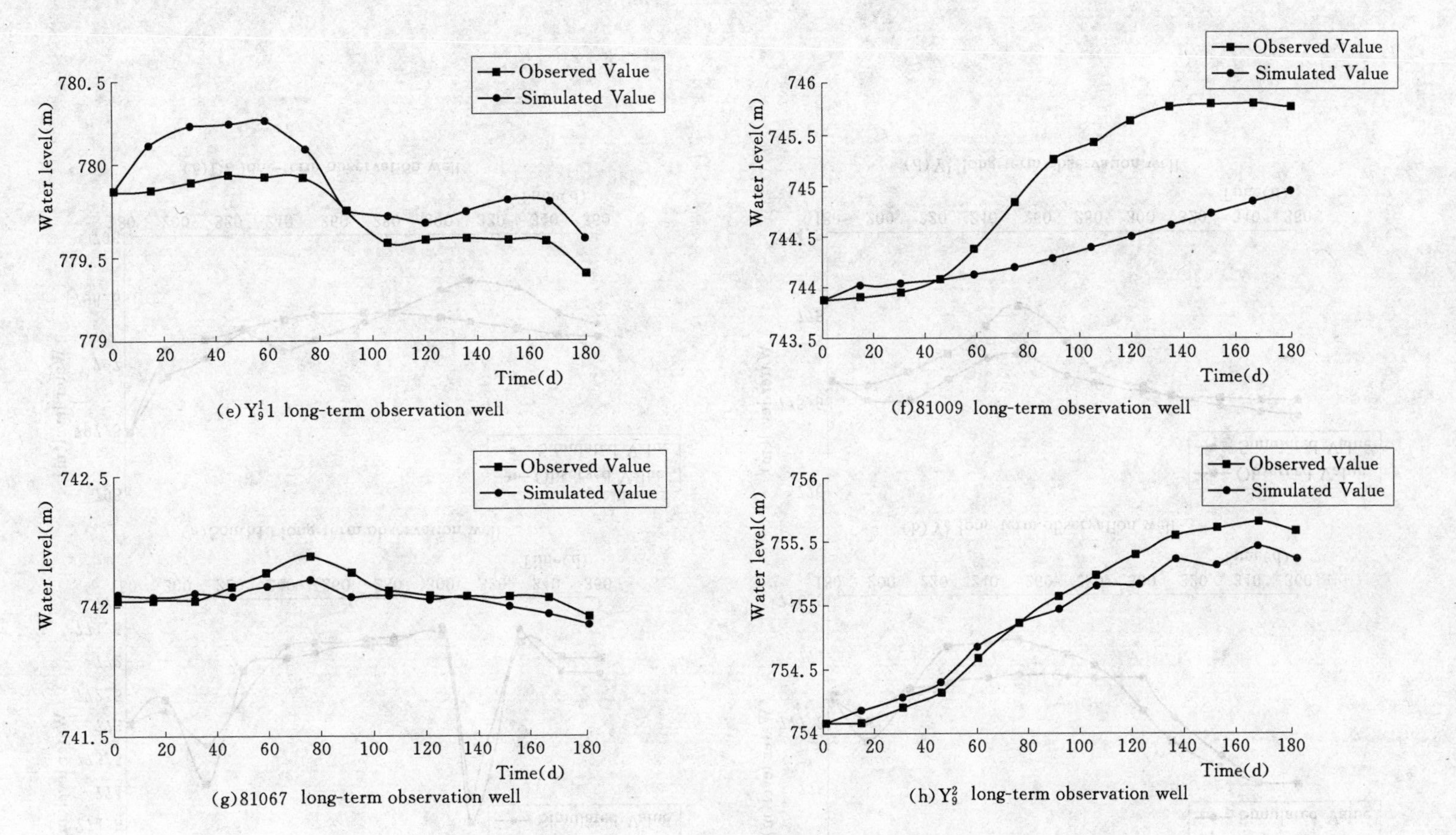

Figure 5.8 Water level fitting curves between the observed head and simulated values

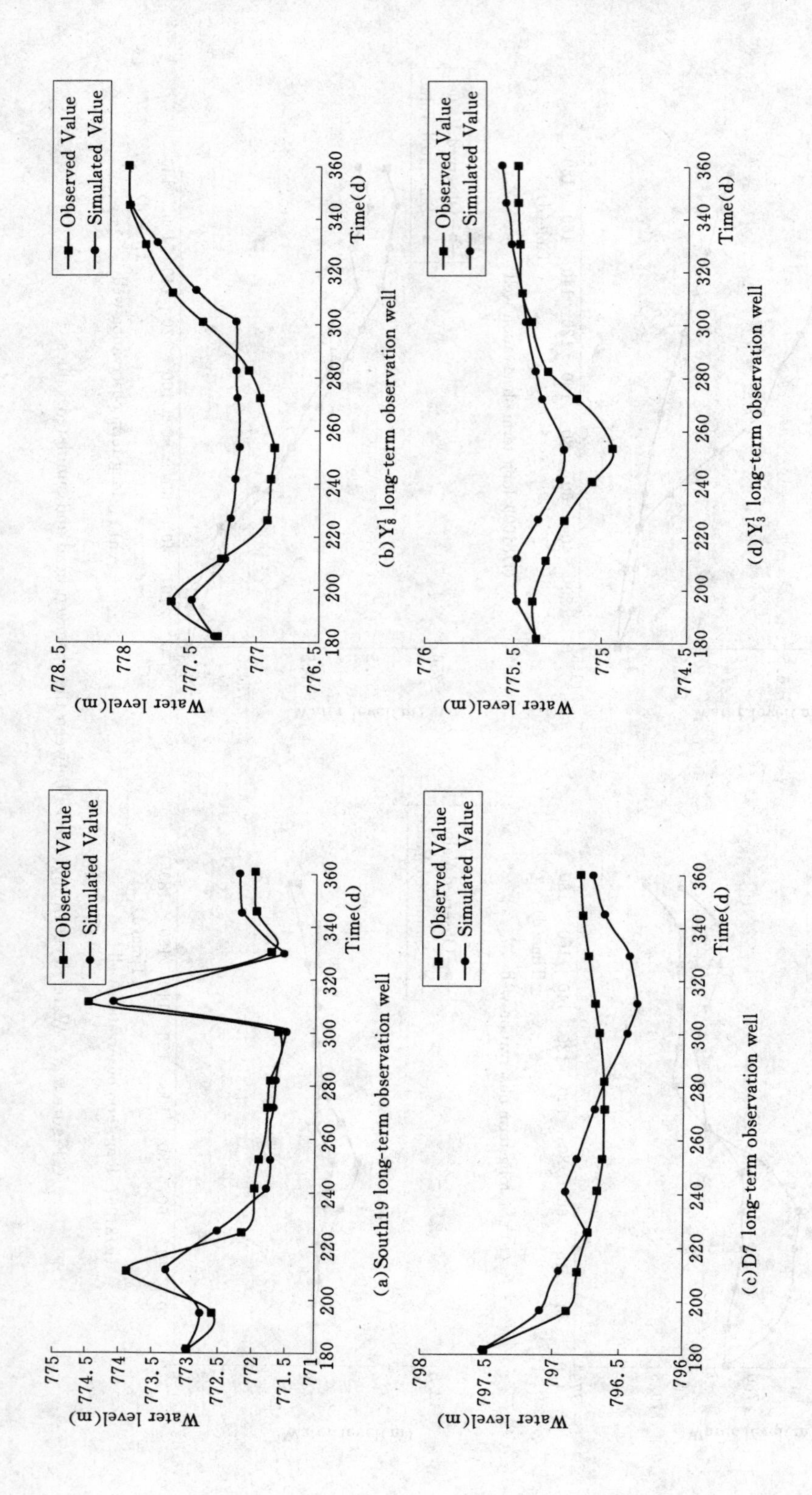

(a) South19 long-term observation well

(b) Y_8^1 long-term observation well

(c) D7 long-term observation well

(d) Y_3^1 long-term observation well

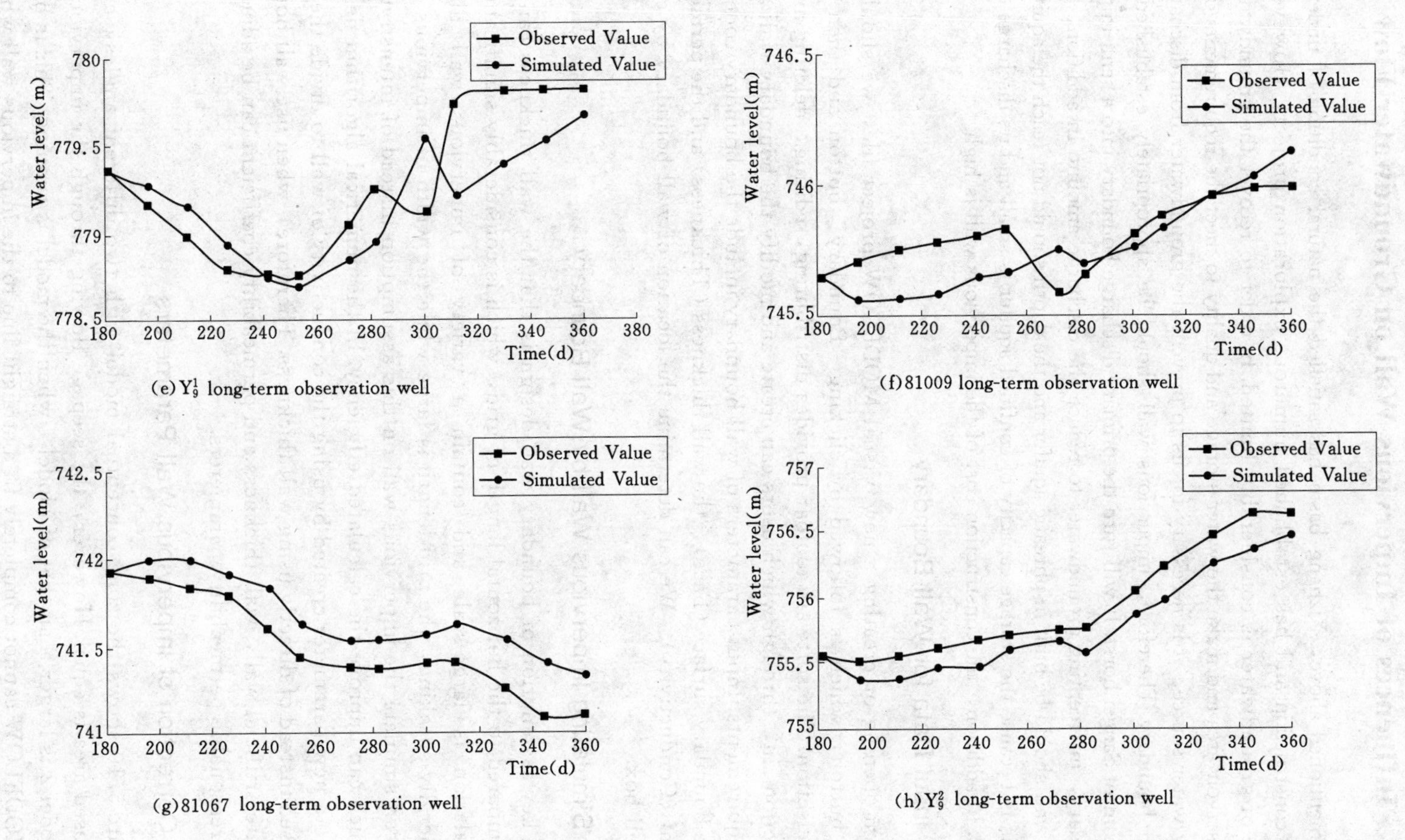

Figure 5. 9 Comparison curves of the observed head and calculated values in the validation period

§ 5.5 Influences of Impervious Wall on Groundwater Flow

As mentioned above, Xizhang basin has satisfied the natural condition of underground reservoir construction and has established numerous exploration projects. However, the recharge of groundwater is considerably limited. In order to recover the groundwater level of the reservoir area and make the water storage and quality to meet the requirements of the emergency water source, it is necessary for us to prevent the groundwater from discharging to the south boundary. Therefore, impervious wall should be appropriately established in the downstream of Sangei horst. It will turn the primary discharge boundary into a semi-impermeable boundary, intercepting groundwater to reduce the discharge amount. In addition, the impervious wall should be built in phreatic aquifer and the depth should not reach the aquitard aquifer. This is because the storage capacity of confined aquifer is small and its discharge capacity is limited. In addition, the construction cost of the impervious wall is high.

5.5.1 Input Data for Wall Boundary

When opening or creating a new Visual MODFLOW project file, we should click "Input" button, which is followed by clicking "Boundary" button and then choose "wall" selection. The software toolbar provides distributing, editing, deleting and copying functions in, through which users can create and modify the boundary wall according to requirements. Input parameters of wall boundary include the boundary code (Code #), the boundary surface (Face), the wall thickness (Thickness) and the permeability coefficient (Conductivity). We can determine the location of wall boundary by clicking on the cell box.

5.5.2 Simulating Impervious Wall by Wall Boundary

The key assumption of boundary wall design is that the wall thickness can be ignored compared to the horizontal length of grid, which is considerably significant when the simulation is large-scale and contain a variety of impervious wall distribution. Generally speaking, the grid length is large while the width of impervious wall is not. We can simulate the impervious wall on this assumption instead of intensifying the grid, which can improve the calculated efficiency. In the numerical algorithm design of software, a new parameter created by using the conductivity of wall to divide the thickness is used instead of directly using wall thickness. Therefore, when use wall boundary to simulate artificial wall, wall thickness and permeability coefficient can be adjusted to obtain a reasonable permeability parameters.

5.5.3 Correction of Impervious Wall Parameters

Figure 5.10 shows the zone conceptual model with two different angles, and the two enclosed area is equal. However, the seepage flowing through the impervious wall in the Model 2 is larger than that of Model 1 when the model is running. This is because Visual MODFLOW cannot completely fit a straight line to the impervious wall when the impervious wall is not parallel to the grid line.

Figure 5. 11 amplify the impervious wall in Model 2. According to Figure 5. 11, we can find that the wall boundary is serrated and the total length of wall boundary will be longer than the impervious wall length in the conceptual model. This means that there is more fluid flowing through the wall boundary and resulting in excessive leakage. To compensate this problem caused by the long wall boundary, the hydraulics parameters must be modified. Therefore, it is proposed a correction method by reducing the permeability coefficient of the wall boundary. Both sides of the right triangle shown as Figure 5. 12 are calculated by the following formula.

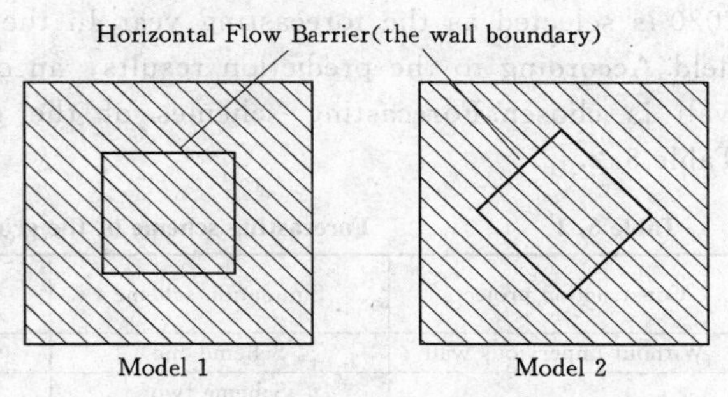

Figure 5. 10 Two zone concept model of impervious wall

$$HYDCHR1_{num\_mod} = HYDCHR_{/su/ls} \times \cos a$$
$$HYDCHR2_{num\_mod} = HYDCHR_{/su/ls} \times \sin a$$

Where: $HYDCHR1_{num\_mod}$ and $HYDCHR2_{num\_mod}$ are the hydraulic parameters of wall boundary respectively; $HYDCHR_{/su/ls}$ is the hydraulic parameter of the actual impervious wall.

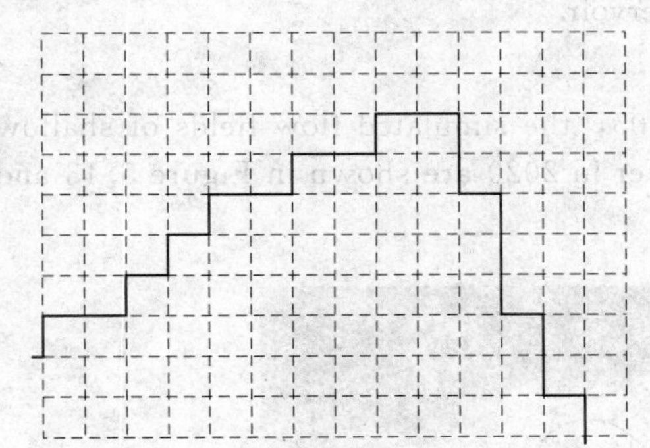

Figure 5. 11 Partial enlarged map about model 2

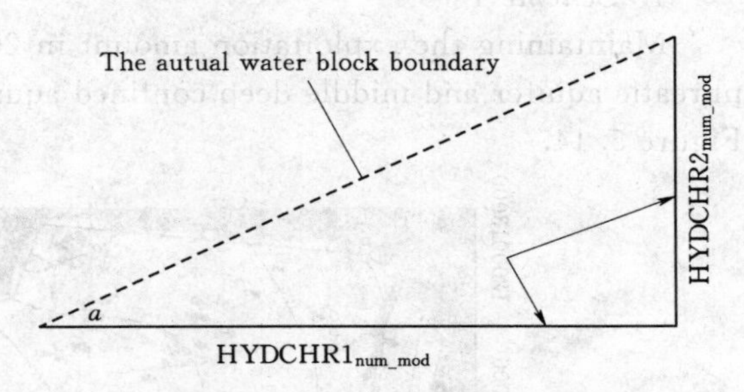

Figure 5. 12 Correction method of hydraulic parameter of the boundary

Each grid can only have one side as the barricade wall during the correcting. Therefore, when a diagonal line is taken as wall boundary, each grid can only set up one side serving as boundary. Then we must edit the grid line, and specify the adjacent grid vertical line to complete the barricade wall boundary. Therefore, the barricade wall boundary consists of the vertical side, which comes from the adjacent unit.

5. 5. 4 Effects of Impervious Wall Construction on the Simulation of Groundwater Flow

The construction scheme of impervious wall can be divide in to three groups, which is shown as Table 5. 2. The thickness of impervious wall has two selections (0. 5m and 0. 15m) while the permeability coefficient can be selected from 0. 01m per day or 0. 001m per day. According to Taiyuan municipal control of groundwater exploration planning,

2020 is selected as the forecasting year. In the above scheme, we predicted the flow field. According to the prediction results, an optimal construction plan of impervious wall is chosen. Forecasting schemes of the groundwater flow field are shown in Table 5. 2.

Table 5. 2 Forecasting scheme of the groundwater flow field

Construction projects	Simulation scheme	The thickness of impervious wall (m)	Permeability coefficient of impervious wall
Without impervious wall	Scheme one		
Impervious wall	Scheme two	0. 5	0. 01
	Scheme three		0. 001
	Scheme four	0. 15	0. 001

After Xizhang underground reservoir storing water, the groundwater flow field will be changed and the groundwater level will have different degrees of recovery, which may cause a series of water environmental problems. In order to quantitatively predict the groundwater level variation characteristics and varying value, and to study the effects of different impervious wall construction schemes on the regional groundwater flow field, we take the groundwater level on January 1, 2005 as the initial flow field. Analysis is conducted by comprising the groundwater flow field of different construction schemes of Xizhang underground reservoir.

1. *Scheme* 1

Maintaining the exploitation amount in 2005, the simulated flow fields of shallow phreatic aquifer and middle deep confined aquifer in 2020 are shown in Figure 5. 13 and Figure 5. 14.

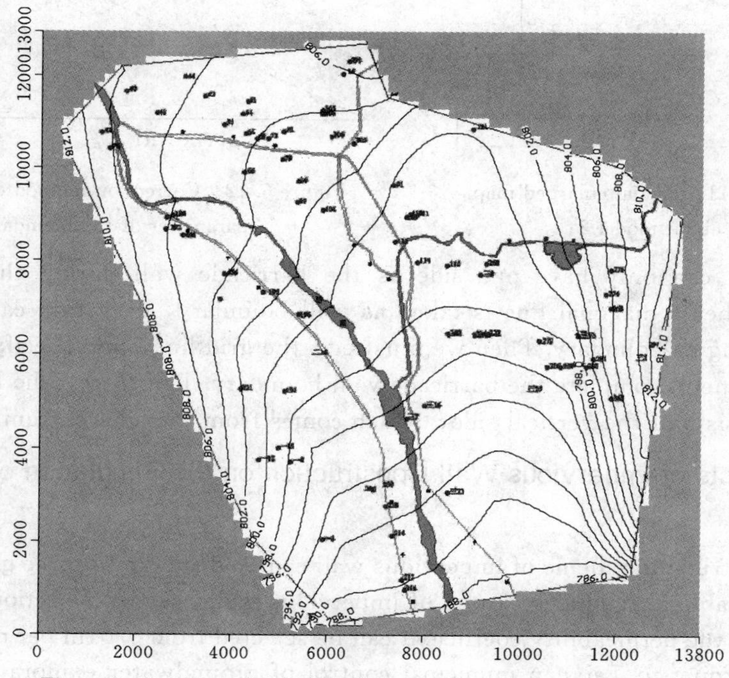

Figure 5. 13 Flow field of shallow phreatic in 2020 (Scheme 1)

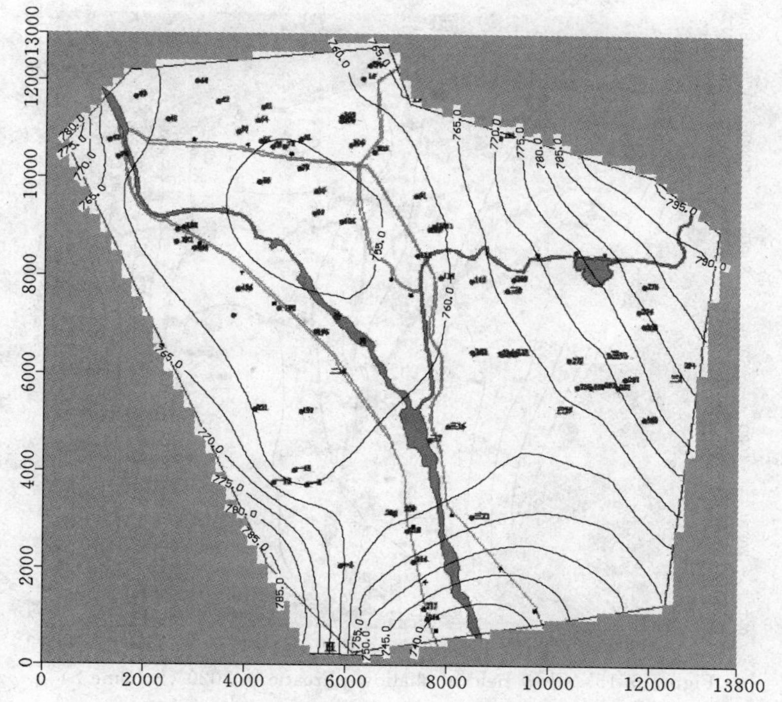

Figure 5. 14 Flow field of confined water in 2020 (Scheme 1)

After 15 years, the groundwater will still flow from north to south in phreatic aquifer on the whole. The groundwater hydraulic gradient of the original depression cone decreases in the simulation area and the water level elevation range from 786m to 814m. The flow field change in confined aquifer will be minor. The increase of water level is between 5m to 15m and the hydraulic gradient will be further decreased.

Based on the above analysis, we find that if only rely on the natural recharge without other measures, the exploitation planning targets of Taiyuan cannot be met, although the center water level of groundwater depression cone will rise in different degrees.

2. *Scheme* 2

Lateral discharge of phreatic aquifer is 8199040m^3 per year if there is no impervious wall in Sangei horst while the lateral runoff discharge will decrease if the impervious wall is constructed. Under the condition of Scheme 2, the simulated flow fields of shallow phreatic aquifer and middle deep confined aquifer in 2020 are shown in Figure 5. 15 and Figure 5. 16. The flow fields will have a small change compared with Scheme one after 15 years. Owing to influence of impervious wall, the hydraulic gradient along with the impervious wall will be reduced in the southern boundary.

3. *Scheme* 3

Under the condition of Scheme 3, the simulated flow fields of shallow phreatic and middle deep confined water in 2020 are shown in Figure 5. 17 and Figure 5. 18. The flow field of shallow phreatic will have a larger change compared with Scheme 1 after 15 years. To illustrate, the hydraulic gradient will be smaller and the lateral discharge be-

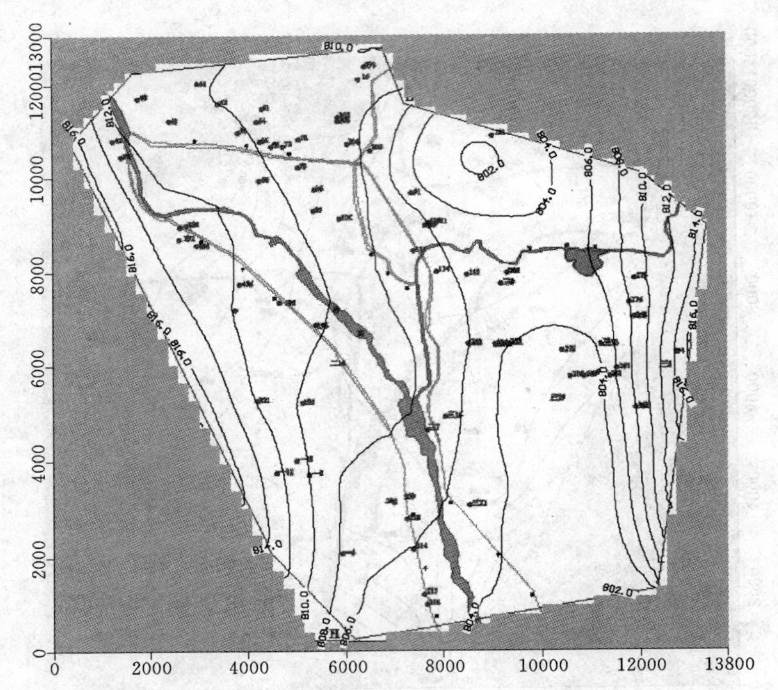

Figure 5. 15 Flow field of shallow phreatic in 2020 (Scheme 2)

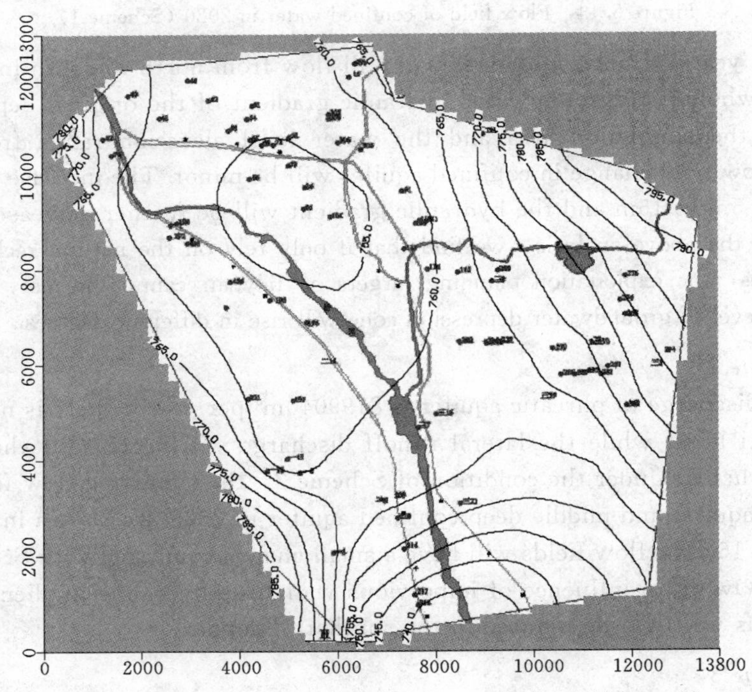

Figure 5. 16 Flow field of confined water in 2020 (Scheme 2)

come less. The construction of impervious wall reduces the discharge and has a great influence on the groundwater flow field. The variation characteristics of confined water flow field are basically the same with Scheme one. Namely the flow field change trend is same and the water level rises a little.

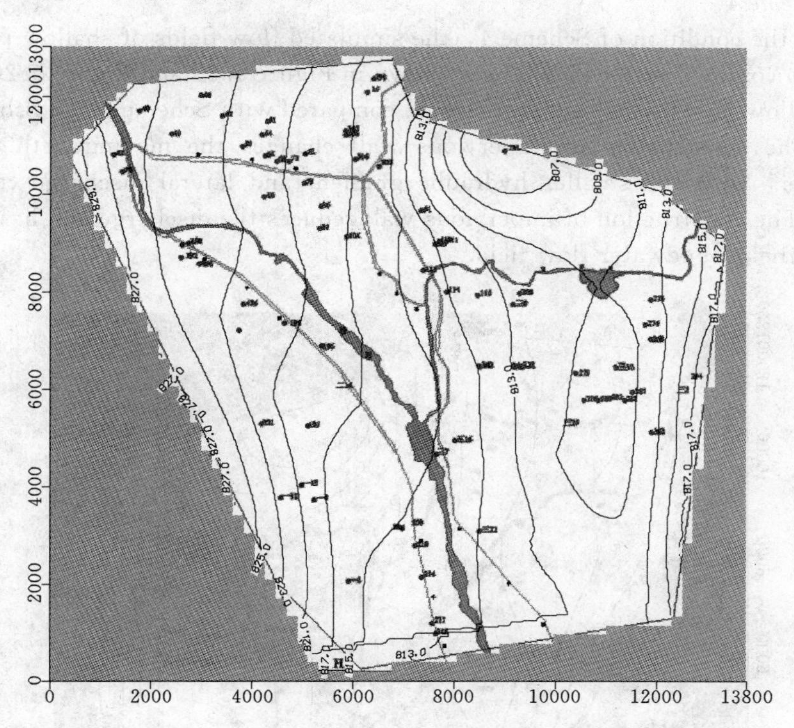

Figure 5.17 Flow field of shallow phreatic in 2020 (Scheme 3)

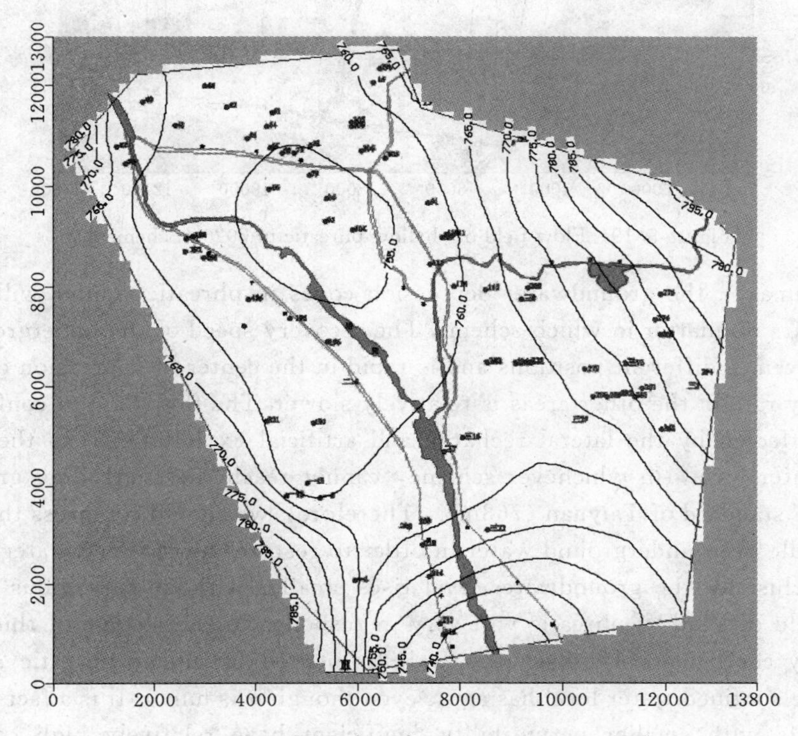

Figure 5.18 Flow field of confined water in 2020 (Scheme 3)

4. *Scheme* 4

Under the condition of Scheme 4, the simulated flow fields of shallow phreatic and middle deep confined water in 2020 are shown in Figure 5. 19 and Figure 5. 20. The flow field of shallow phreatic has a larger change compared with Scheme 2 and Scheme 3 after 15 years. The construction of impervious wall changes the moving path of regional groundwater, with the smaller hydraulic gradient and lateral discharge compared to Scheme 1. The construction of impervious wall reduces the discharge and has a great influence on the groundwater flow field.

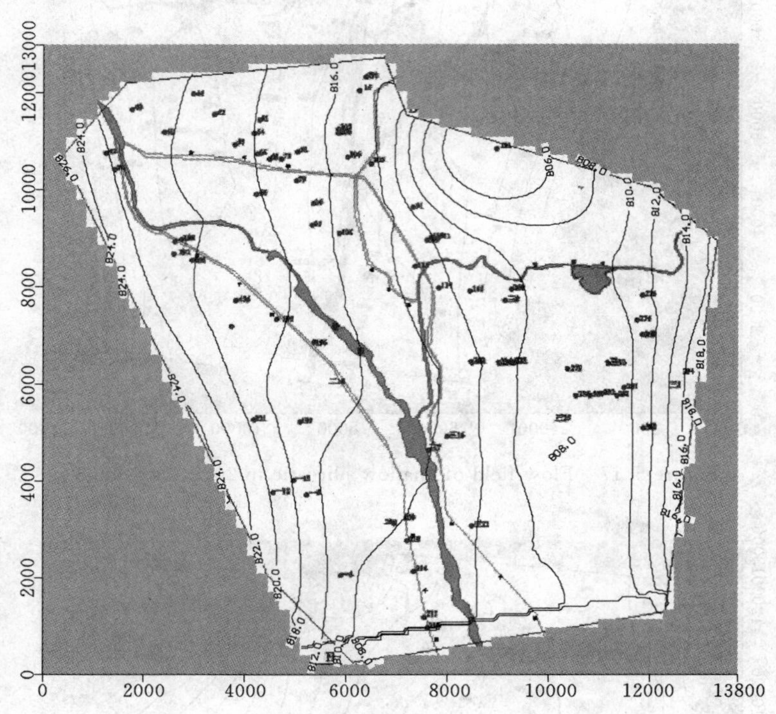

Figure 5. 19 Flow field of shallow phreatic in 2020 (Scheme 4)

In summary, the groundwater depression cones in phreatic aquifer will disappear after 15 years no matter in which scheme. The recovery speed of the underground water level is uneven in different positions and is rapid in the center of depression cones while the recovery rate in the other areas is relatively slower. The flow field of confined water is mainly affected by the lateral recharge and artificial exploitation, so the recovered confined water level, in whichever scheme, cannot reach the controlling groundwater exploitation standard of Taiyuan (768m). Therefore, we should compress the exploitation in middle deep underground water in order to restore the confined water level.

In conclusion, the groundwater level rises smaller without impervious wall while the flow field of shallow phreatic will vary in response to the change of thickness and permeability coefficient of impervious wall. Influenced by upper phreatic aquifer recharge, the confined water level has a rise even though it is minor. It is a fact that building materials with smaller permeability coefficient have relatively high construction

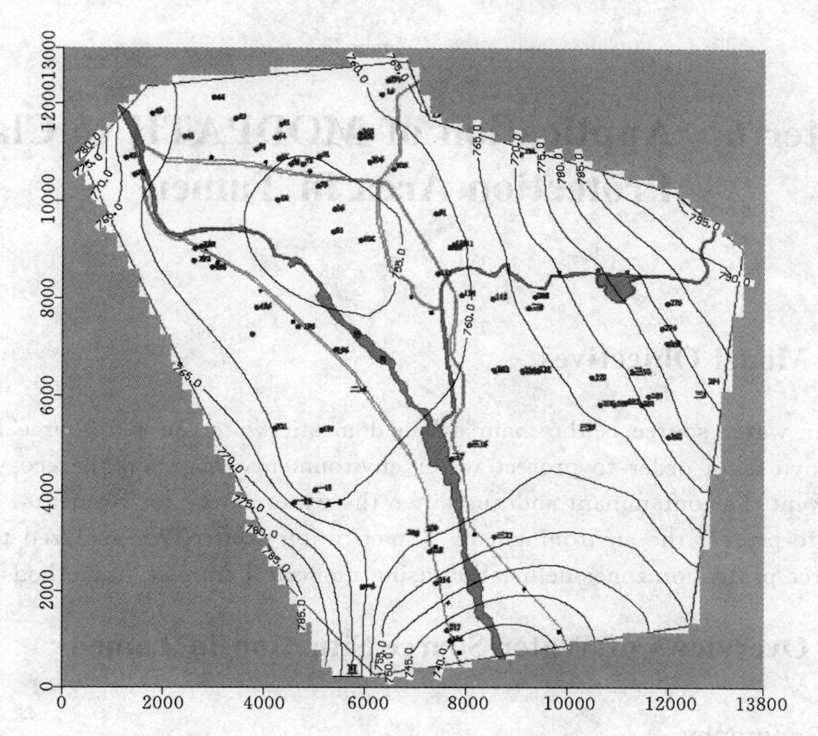

Figure 5. 20 Flow field of confined water in 2020 (Scheme 4)

cost. Therefore, in consideration of the economic factor and the increase of underground water level, we choose the impervious wall with a thickness of 0. 5m, permeability coefficient 0. 01m per day as an optimal construction scheme.

Chapter 6 Application of MODPATH to Classify Protection Area in Tumen

§ 6.1 Model Objectives

Tumen water source is the main urban domestic water supply source in Linfen, Shanxi Province. In order to protect water environment, maintain the ecosystem balance, prevent the contaminant and guarantee the water safety for residents, it is quite necessary to protect the environment of Tumen water source. We analyzed the Tumen water source protection zones delineation using numerical simulation methods.

§ 6.2 Overviews of Water Source Situation in Tumen

6.2.1 Geography

1. *Location and topography*

Tumen water source area lies in the west of Fen River in Linfen, 15km to the south of Tumen town. The water source area is about 7000m from north to south and 2300m from east to west. The shape of the research area is a parallelogram, with a total area of about 15.8km^2 and the exploitation area of 10.7km^2. Tumen water source area has a beautiful environment and convenient transportation. The location and traffic conditions of Tumen water source is shown in Figure 6.1.

Tumen water source area is a large alluvial skirt which is made up of four alluvial fans. The area can be described as undulating terrain and sloping terrain towards the basin. Piedmont alluvial fan tilts to the plain which is higher at northwest but lower at southeast. The gradient of the ground varies from 4‰ to 10‰ while the elevation from 500m to 600m. The western mountain has marked uplifts and the piedmont is mainly with thick gravel and fallen broken boulders (maximum size of 2m to 3m). The gradient of top fan is 7‰ to 10‰. The lower and front part of alluvial fan is of gravel or sand with the gradient of 4‰ to 6‰. The west side of the fault zone is Lvliang Mountain with elevation of 800m to 1400m while the eastern side is connected to Linfen basin. The karst fissures of the fracture zone and the valley contribute to atmospheric precipitation infiltration to the groundwater and groundwater runoff generation.

2. *Meteorology and hydrology*

Tumen Water source area has the features of a continental arid and semi-arid climate. It can be described as cold and dry in winter while hot and rainless in summer. In

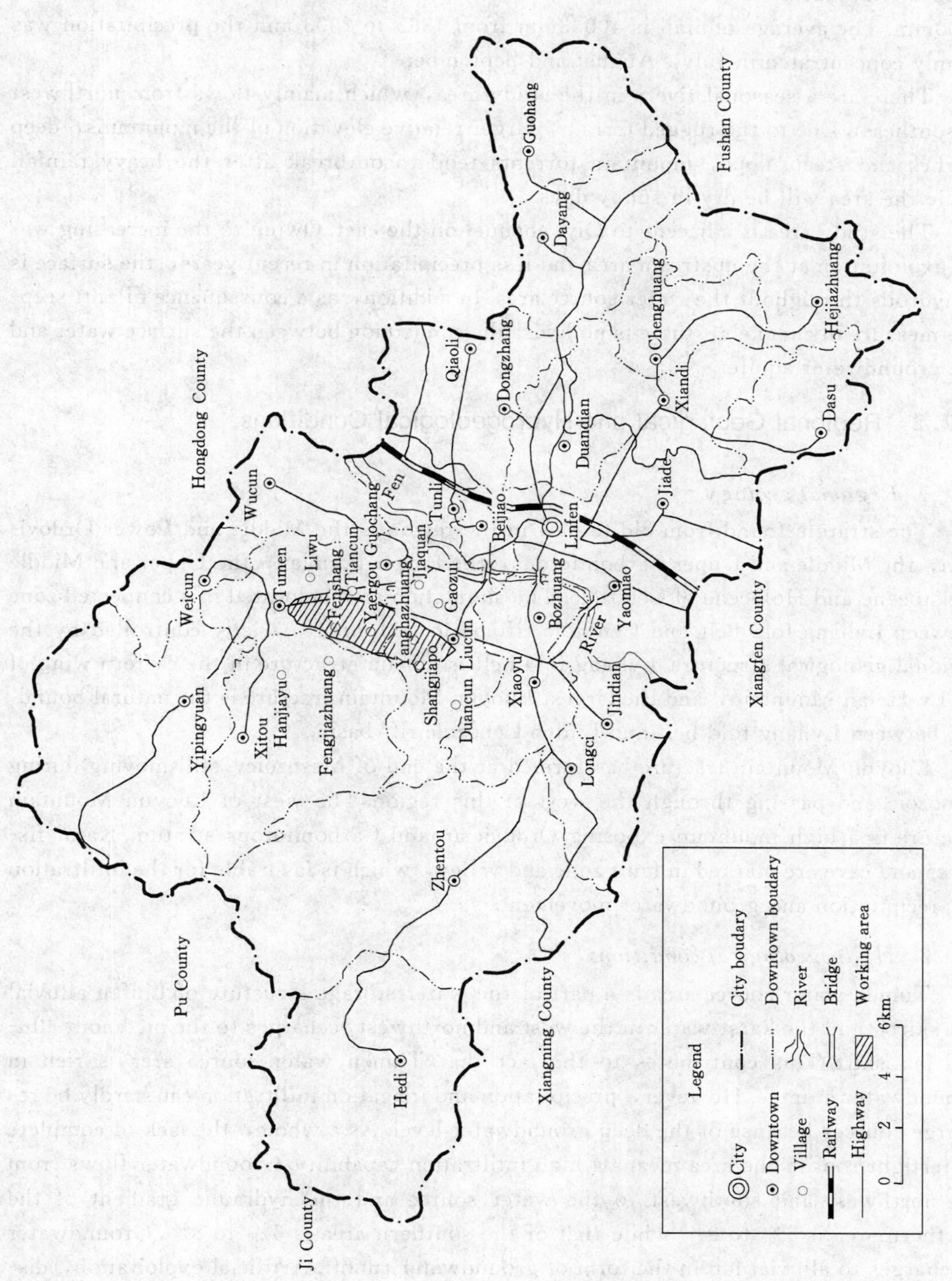

Figure 6.1 Location and traffic conditions of Tumen water source area

addition, draught happened in the most time of year, especially in spring. The average annual temperature is 12.1℃, and the annual evaporation varies from 1800mm to 2000mm. The average rainfall is 400.3mm from 1965 to 2003 and the precipitation was mainly concentrated in July, August and September.

There are 4 seasonal rivers in the study area, which mainly flows from northwest to southeast. Due to the rugged terrain, larger relative elevation of the mountains, deep ditches and steep slope, mountain torrents tend to outbreak after the heavy rainfall while the area will be dry in sunny days.

The study area is adjacent to Qiyi channel on the east. Owing to the increasing water exploitation at the upstream area and less precipitation in recent years, the surface is anhydrous throughout the water source area. In addition, as a consequence of anti-seepage measure in channels, there is no hydraulic connection between the surface water and the groundwater aquifer.

6.2.2 Regional Geological and Hydrogeological Conditions

1. *Regional geology*

The stratum found from old to new in the region is the Middle and Lower Ordovician, the Middle and Upper Carboniferous, the Lower Permian, the Upper and Middle Pleistocene and Holocene of Cenozoic. The simulation area is located in a connected zone between Lvliang fold belt and Cenozoic rift basin in Linfen, strictly controlled by the regional geological structure. Lvliang fold belt is a main structure in the eastern wing of Qi-Lv-Helan Mountain, and the largest Luoyun Mountain fracture is the natural boundary between Lvliang fold belt and Linfen Cenozoic rift basin.

Luoyun Mountain fracture has formed at the end of Mesozoic, still moving during Cenozoic and passing through the west of this region. The west of Luoyun Mountain fracture is a high mountain exposing Ordovician and Carboniferous stratum. Karst fissures and cave are marked in fault zone and valley, which is favorable for the infiltration of precipitation and groundwater movement.

2. *Hydrogeological conditions*

Tumen water source area is a part of the water storage structure of Linfen alluvial fan skirt, and the karst water in the west and northwest recharges to the piedmont alluvial fan skirt. This contributes to the fact that Tumen water source area is rich in groundwater storage. However, precipitation and irrigation infiltration can hardly be recharge sources because of the deep groundwater level. As a whole, the lack of complete watertight roof in the area means a high infiltration capability. Groundwater flows from the northwest and southwest to the water source and the hydraulic gradient of the northern area is 6‰ to 9‰ while that of the southern area is 5‰ to 8‰. Groundwater discharges to alluvial fan in the form of groundwater runoff. Artificial exploitation, discharge in the front of alluvial fan in the form of spring and evaporation is the main discharge way.

§ 6. 3 Groundwater Flow Model for Tumen Water Source Area

6. 3. 1 Hydrogeological Conceptual Model

The groundwater system is piedmont alluvial fan skirt, and the aquifer ranges from single layer to multi layers from the fan head to the front. Correspondingly, the lithology ranges from gravel, sand and gravel to medium sand and fine sand. According to the pumping test and drill data, the west of simulation area is unconfined water while the east is confined water, with a uniform hydraulic flow field for both areas. At present, the withdraw wells are all of mixed exploitation of the unconfined water and confined water. Therefore, we regard the groundwater system as heterogeneity isotropous and mixed aquifers, and establish an unsteady two-dimensional groundwater flow model for the unconfined and confined aquifer.

The eastern and western boundary is the lateral recharge and discharge boundary respectively. The boundaries can be generalized as constant flow boundary because the quantity of recharge and discharge is known. We regard the recharge and discharge boundary as injection and pumping wells respectively in the simulation. The northern and southern boundaries both are low permeability boundaries, and thus can be generalized as impermeable boundaries. Meanwhile, flow line is considered to be a zero flow boundary because the northern and southern boundary is divided by it.

6. 3. 2 Mathematical Model for the Groundwater Flow

1. *Establishment of the mathematical model*

Based on the hydrogeological conceptual model, a mathematical model for unsteady 2-D flow in heterogeneity isotropic aquifers is established. According to the Dupuit assumption, the mathematical model is shown as following:

$$\begin{cases} \frac{\partial}{\partial x}\left(kh\frac{\partial H}{\partial x}\right)+\frac{\partial}{\partial y}\left(kh\frac{\partial H}{\partial y}\right)+W=\mu\frac{\partial H}{\partial t}+\sum_{i=1}^{n}Q_i\delta(x-x_i,y-y_i) & (x,y)\in D,t>0 \\ H(x,y,t)\big|_{t=0}=H_0(x,y) \quad (x,y)\in D,t>0 \\ kh\frac{\partial H}{\partial n}\Big|_{\Gamma_1}=q_1(x,y,t) \quad (x,y)\in\Gamma_1,t>0 \\ kh\frac{\partial H}{\partial n}\Big|_{\Gamma_2}=0 \quad (x,y)\in\Gamma_2,t>0 \end{cases} \tag{6.1}$$

Where: H is groundwater head, [L]; k is the permeability coefficient of aquifer, [LT^{-1}]; h is the thickness of aquifer, [L]; μ is the specific yield; Q_i stands for the total discharge of pumping well i; n is the number of pumping wells; $\delta(x-x_i,y-y_i)$ is a function of unit well; (x,y) is the expression of space rectangular coordinates, [L]; t is the time, [T]; W is vertical exchange quantity, [LT^{-1}]; D stands for the study area; $H_0(x,y)$ is the initial head of study area, [L]; Γ_1 represents locations of western and eastern boundaries; Γ_2 represents locations of the second type boundaries;

$q_1(x,y,t)$ is the function of flow distribution at the second type boundary, $[L^2T^{-1}]$; n is the outside normal vector of the seepage boundary.

Visual MODFLOW, which is the most popular for practical applications in three-dimensional groundwater flow, is applied to solve the mathematical model.

2. *Grid subdivision*

The area of the simulation region is 10.7km^2. The simulation region is divided into 47 rows and 25 columns with the 1175 cells totally. According to the actual topography, 971 cells are considered to be active while the others are regarded as inactive. The result of grid subdivision is showed in Figure 6.2.

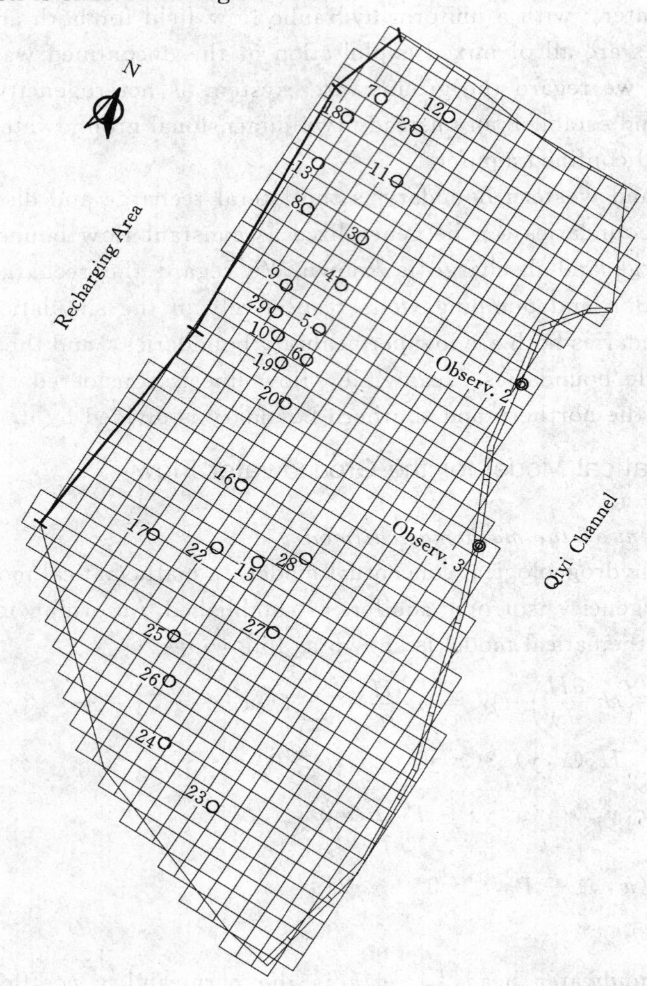

Figure 6.2　Grid subdivision of the simulation area

3. *Boundaries and initial condition*

Boundaries condition. Based on the large number of hydrogeology data and the initial flow field, the boundaries are determined as following:

The western boundary controlled by the piedmont fault, can be considered as a lateral recharge boundary. The eastern boundary is lateral recharge boundary with known discharge amount. Therefore, the western and eastern boundary can be seen as second boundary, which is called as known-flow boundary. The north and south is impervious boundary delimited along with groundwater flow direction.

Initial condition. As shown in Figure 6. 3, the groundwater contour on April 10, 2000 is selected as the initial flow field.

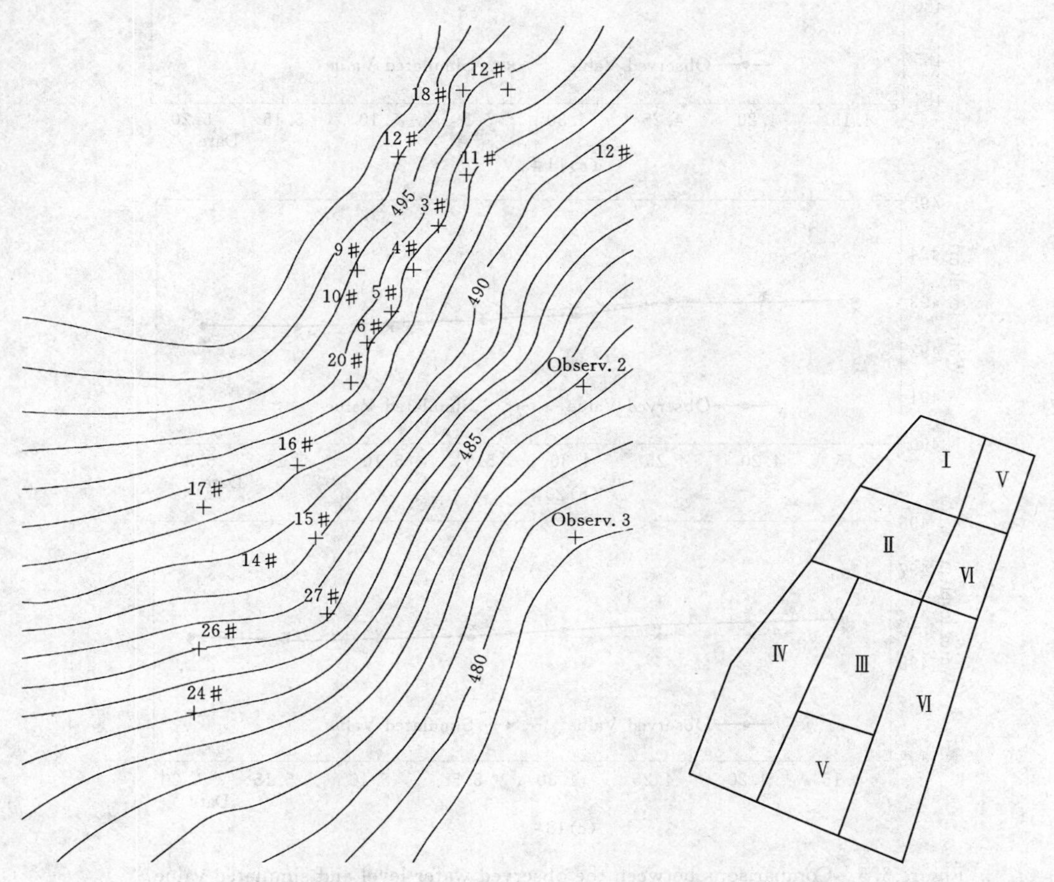

Figure 6. 3 Two-dimension flow field of Tumen groundwater water source area

Figure 6. 4 Hydrogeological parameters partition of Tumen water source area

6. 3. 3 Model Validation

Through adjusting the parameters repeatedly through comparing the flow field and long-time observation water level in the same period, the hydrogeological parameters and the model structure are determined.

Through the parameter identification of six observation holes in eight parameters partition zone during the pumping test (in eight periods from April 10 to May 21, 2000), the hydrogeological parameters partitions are adjusted and shown in Figure 6. 4. Table 6. 1 gives optimized hydrogeological parameters while Figure 6. 5 demonstrates the comparison between the observed water level and simulation.

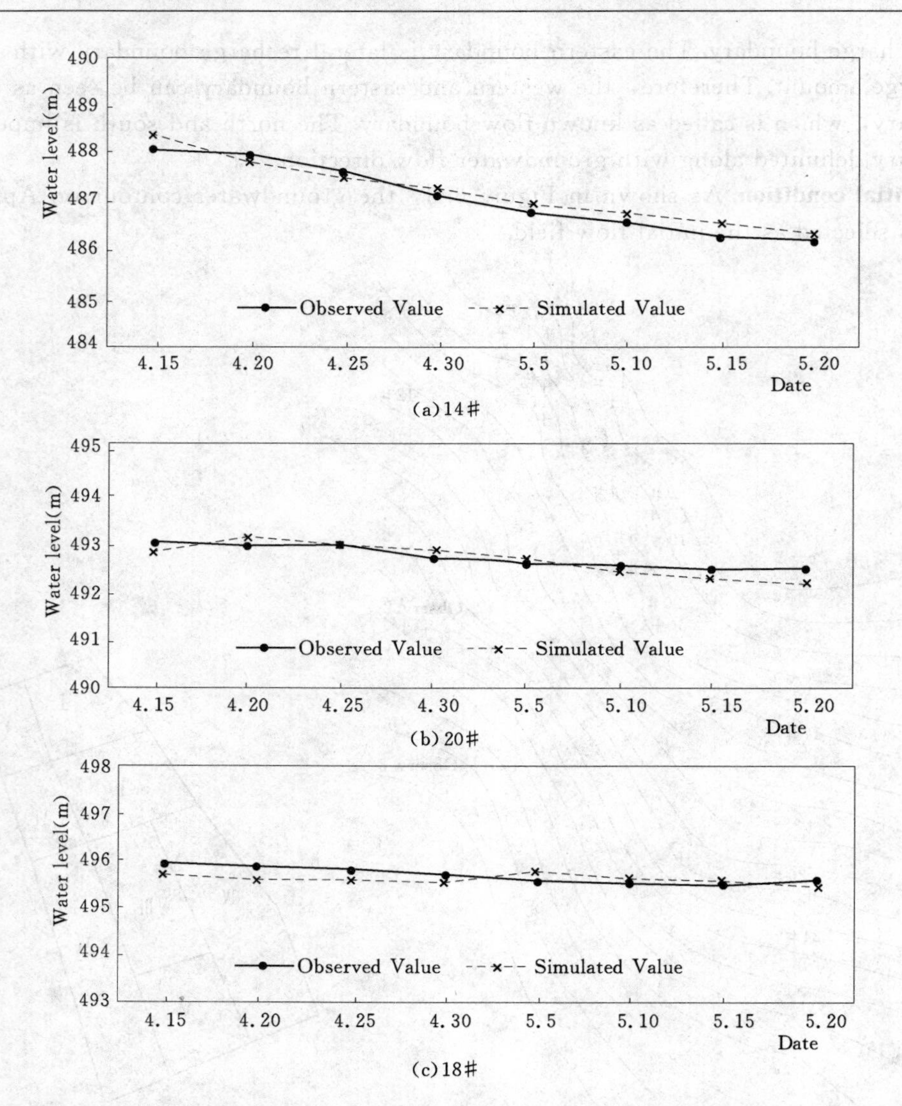

Figure 6.5 Comparisons between the observed water level and simulated value

Table 6.1 Optimized Hydrogeological Parameters

Partition Zone	K(m/d)	T(m^2/d)	μ	Note!
Ⅰ	17.72	806.7	0.256	T takes the average value during the simulation period
Ⅱ	53.12	3622.8	0.273	
Ⅲ	13.21	721.3	0.175	
Ⅳ	4.85	247.4	0.161	
Ⅴ	3.75	18.8	0.122	
Ⅵ	3.18	159.3	0.110	
Ⅶ	3.55	175.0	0.103	
Ⅷ	3.02	120.8	0.098	

In conclusion, water level simulation errors are minor at most time. The errors of simulated water level and observed value are shown in Table 6. 2. 87. 5% of water level errors are less than 0. 3m while 12. 5 % errors are between 0. 3~0. 5m, which indicates that the model is reasonable in structure with reliable parameters. Therefore, the model can be used for simulating groundwater flow field in different periods.

Table 6. 2 Error of simulated water level and observed value

Node number \ Time	1 (4.15)	2 (4.20)	3 (4.25)	4 (4.30)	5 (5.5)	6 (5.10)	7 (5.15)	8 (5.20)
18#	0.282	−0.181	−0.102	0.107	0.205	0.249	0.301	0.18
20#	−0.185	0.227	0.082	0.208	0.066	−0.145	−0.213	−0.415
14#	−0.212	−0.325	−0.239	−0.154	0.182	0.144	0.065	−0.212

§ 6.4 Delineation of Water Source Protection Zones by MODPATH

The technique of Waterpower Interception and Capture (WIC) is widely used for protection zones delineation in foreign countries. WIC zone belongs to a part of the aquifer. For a three-dimensional flow, *t* years WIC zone is the volume surrounded by the path line within *t* years. If ignoring the vertical velocity component, the WIC zone will become a plane area of horizontal line path within *t* years.

6.4.1 Numerical Simulation Method

Based on the groundwater flow field simulated by MODFLOW and time standard of different protective zone, MODPATH with the particle backward tracing package is applied to determine the hydraulic capture zone in different time periods. According to geography and hydrogeological condition of the water source, the protection zones can be delineated.

The three-dimensional particle tracking simulation of unsteady flow in a given time is accomplished by MODPATH. Using groundwater numerical simulation results, MODPATH can calculate the three-dimensional streamline distribution and the water particle location at any time. MODPATH can conduct forward and backward tracking after the particles location is specified. The so-called forward tracking is defined as the tracking particles in the groundwater system supply area, and then tracking water flows from the supply area to the discharge zone; while the backward tracking is defined as the tracking particles in the discharge area, and tracking water flows from the discharge area to the supply zone. The streamline tracing function of MODPATH, especially the backward tracking technique, may directly determine the groundwater recharge sources and supply channels. Meanwhile, it can figure out the time span the groundwater need to flow from the recharge area to the specified study area.

For a three-dimensional steady flow, the mass balance equation of MODPATH can be expressed as following equation:

$$\frac{\partial(nV_x)}{\partial x}+\frac{\partial(nV_y)}{\partial y}+\frac{\partial(nV_z)}{\partial z}=W \tag{6.2}$$

Where: V_x、V_y、V_z is the component of velocity vector in direction of the axis respectively; n is the effective porosity of the aquifer; W is the water yields per unit volume generated by the sources and sinks coverage.

Equation (6.2) can be solved by the finite difference method.

6.4.2 Procedure of Protection Zones Delineation

According to the purposes, principles and time standards of protection zones delineation, comprehensive simulation method including flow field simulation, path line tracking and analysis of the hydrogeology conditions is conducted to delineate the different protection zones. The delineating steps are shown as following:

Step 1 Determining the initial time and the groundwater flow field, and then simulate the flow field distribution from the initial time to 60 days, 10 years and 25 years later.

Step 2 Determining the WIC zones on 60 days, 10 years and 25 years later by the WIC technique.

Step 3 Based on the hydrogeological conditions and above simulation results, the protection zones will be delineated finally.

6.4.3 Results of Protection Zones Delineation

1. *Determining groundwater flow field*

A group of wells in the simulation have been conducted pumping test with yield of 39800m^3 per day from April 10 to May 21, 2000. The groundwater flow field after this is shown as Figure 6.6. The figure shows that two larger depression zones have not achieved stability after 42 days continuous pumping with exploitation quantity of 39800m^3 per day. The results show that the exploitation of 39800m^3 per day is greater than the recharged quantity, and thus a series of water environmental problems will be caused by this. In order to avoid large drawdown of groundwater level and destroying groundwater resources, and achieving the sustainable development of groundwater resources utilization, we select the 95% of actual groundwater exploration amount (37800m^3 per day) to simulate.

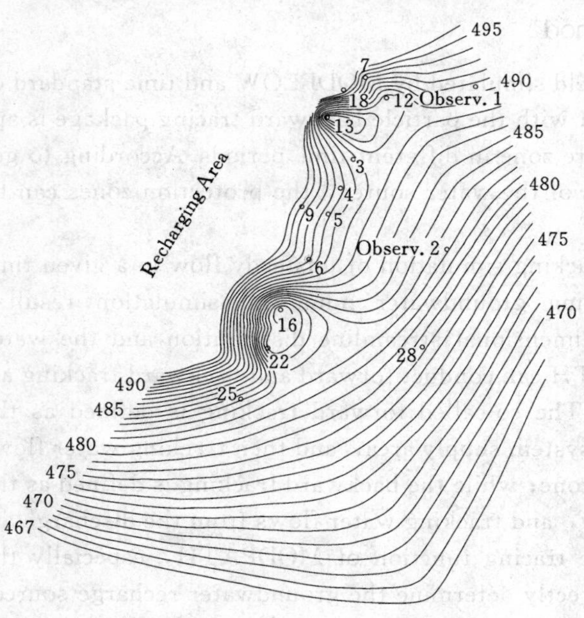

Figure 6.6 Groundwater flow field after pumping test

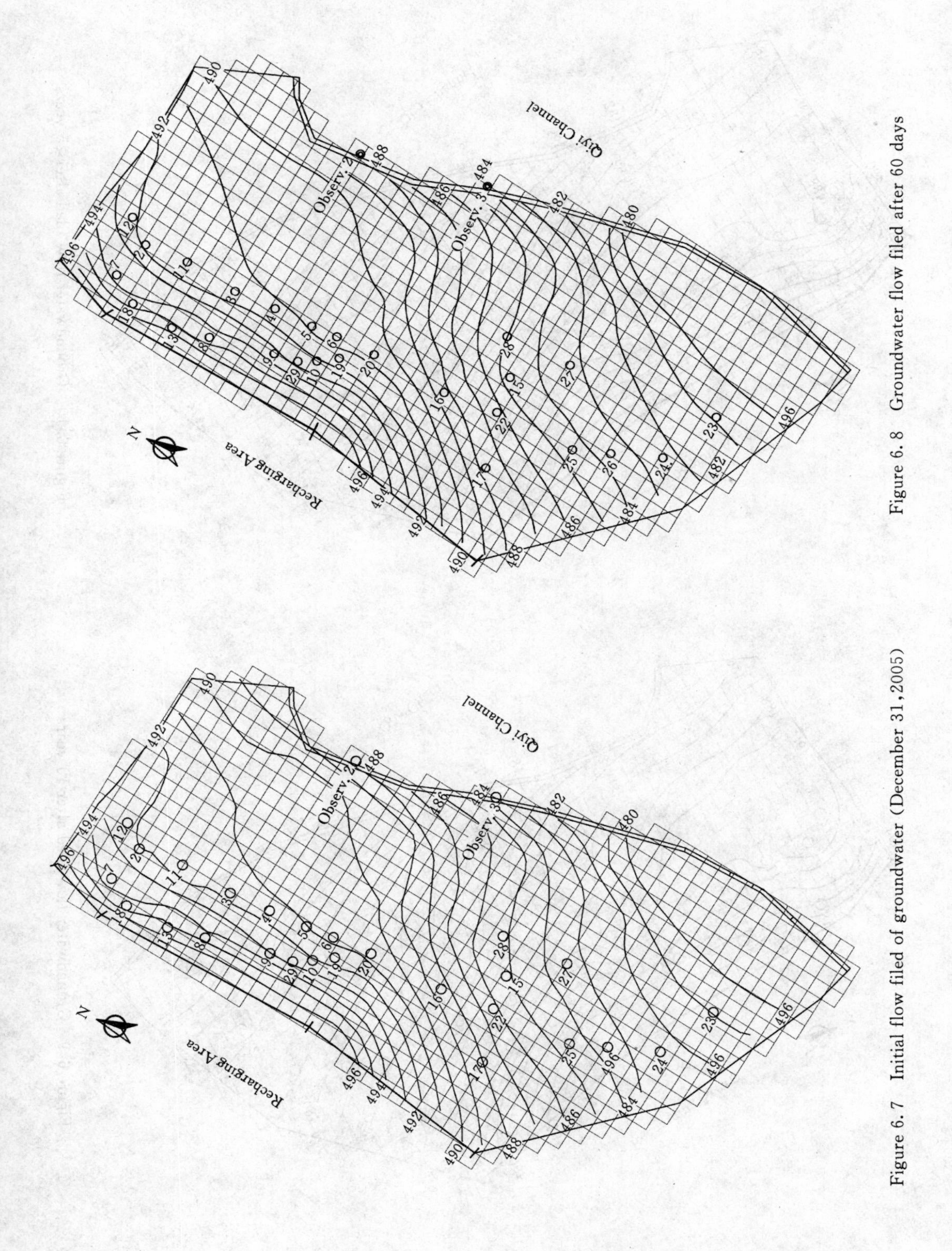

Figure 6.7 Initial flow filed of groundwater (December 31, 2005)

Figure 6.8 Groundwater flow filed after 60 days

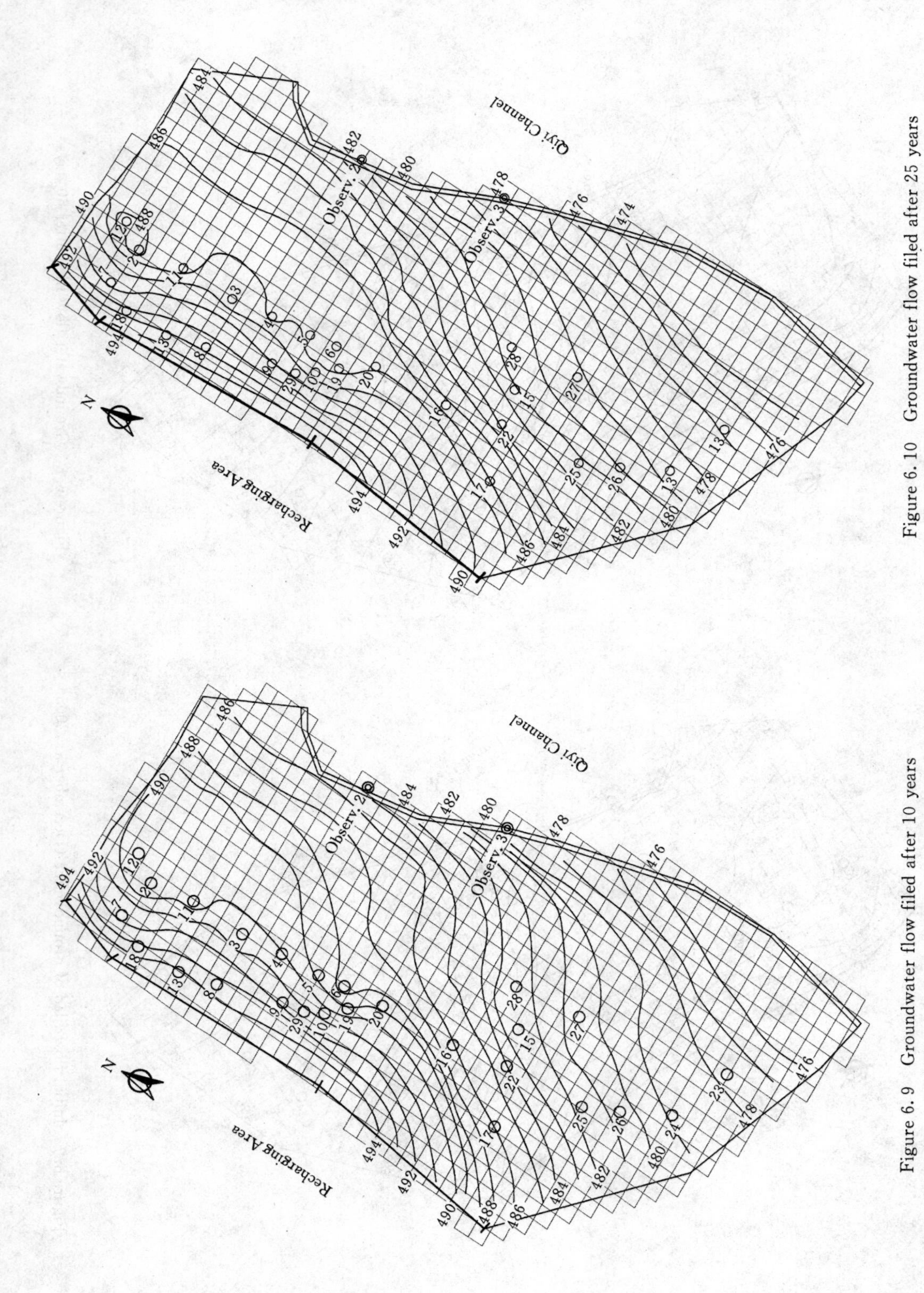

Figure 6.9 Groundwater flow filed after 10 years

Figure 6.10 Groundwater flow filed after 25 years

Taking the simulated result on December 31, 2005 as the initial flow field (Figure 6.7), we simulate the groundwater flow field after 60 days, 10 years and 25 years. The results are shown as Figure 6.8, Figure 6.9 and Figure 6.10 respectively.

2. *Determining the WIC zone*

The WIC zone of different period is acquired by WIC technique tracking backward through the simulation in withdrawal well (Figure 6.11). According to the withdrawal wells distribution and hydrogeological condition, we select eleven withdrawal wells (2#, 3#, 4#, 5#, 6#, 7#, 13#, 15#, 16#, 18#, 22#) that lie in the area rich of water as the production well to place pollution sources, (i. e., simulated tracer particles). Then, we simulate the groundwater backward tracing path line after 60 days, 10 years and 25 years.

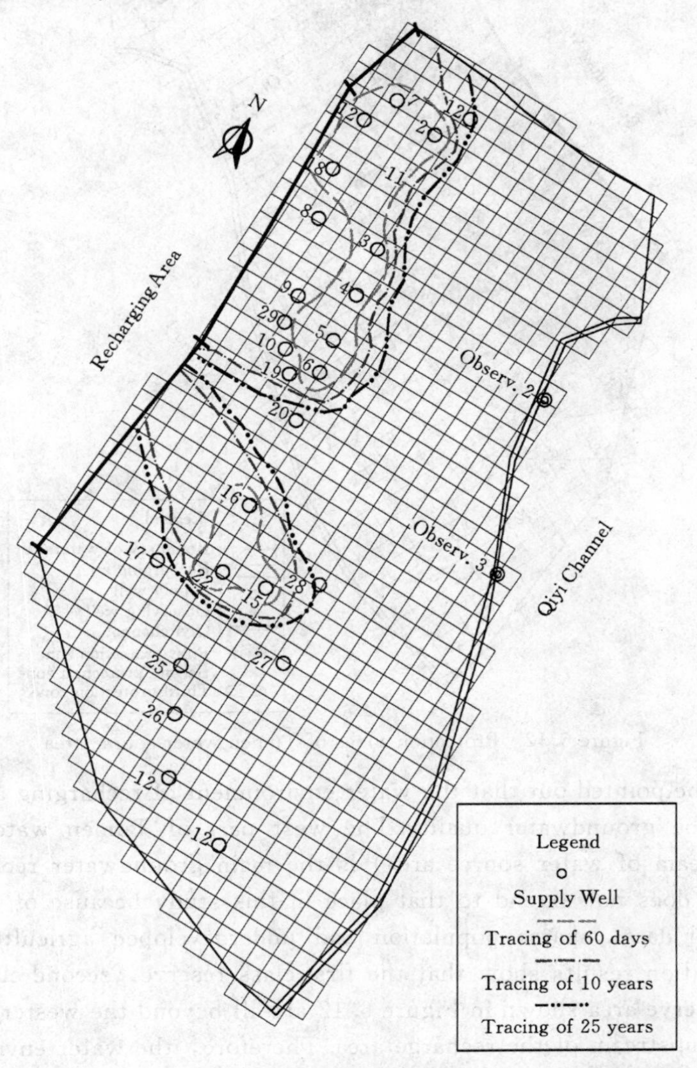

Figure 6.11 WIC zone of different period simulated by MODPATH

3. *Delineation of the protective zone*

Based on above simulation results of flow field and the WIC zone, combined with the withdrawal wells distribution and hydrogeological condition, the protection zones of Tumen Water source area are delineated as shown in Figure 6. 12.

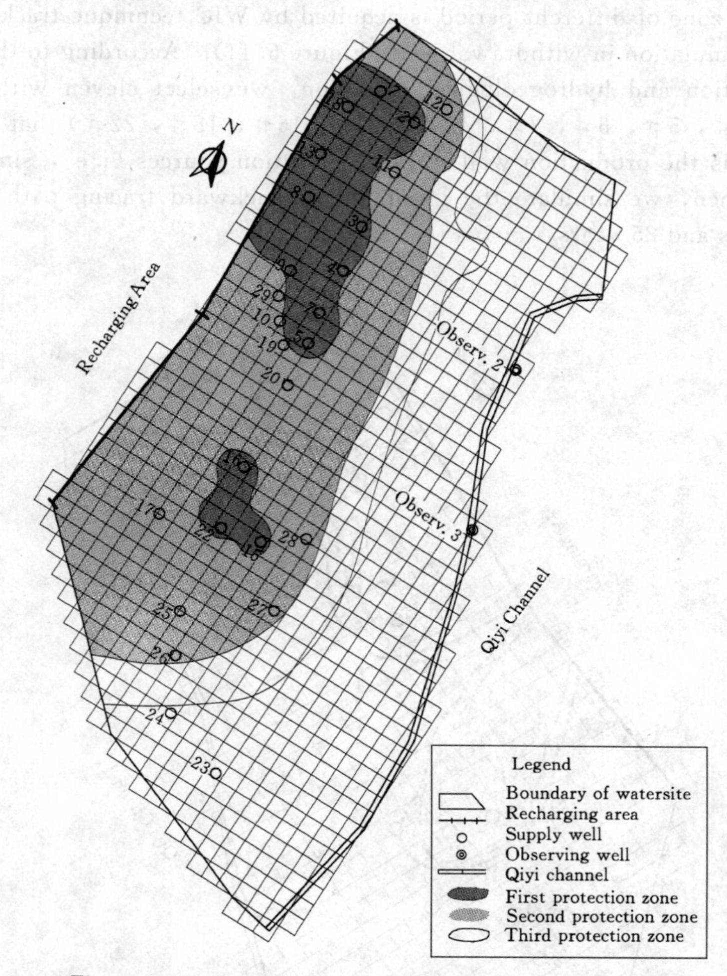

Figure 6. 12 Protection zones of Tumen water source area

It should be pointed out that the water environment of recharging area has a great influence on the groundwater quality. The west part of Tumen water sources area (i. e. the upstream of water source area) is the main groundwater recharge area. The simulated area does not extend to that place in this study because of the shortage of hydrogeological data, sparse population and underdeveloped agriculture and industry. The simulation results show that the first-class reserve, second-class reserve and prospective reserve area shown in Figure 6. 12 are all beyond the western boundary and expand to the upstream of the recharge area. Therefore, the water environment condition of the recharging area must be adequately considered and the corresponding measures should be taken to prevent the water pollution.

References

[1] Singh, S. K. Review of selected softwares for groundwater flow modeling
Hydrology, 1997, (11): 1-13.

[2] 钱会，王毅颖，宋秀玲．地下水流数值模拟中不应忽视的几个工作程序［J］.
2004, (1): 40-43.

[3] 薛禹群，吴吉春．地下水数值模拟在我国——回顾与展望［J］．水文地质工程地质，1
21-24.

[4] 韩再生．地下水资源数值法计算技术要求——行业标准介绍［J］．水文地质工程地质，19
(4): 47-50.

[5] Anderson M P, Woessner W W. Applied groundwater modeling—Simulation of Flow and advective transport [M]. NewYork: Academic Press Inc., 1992.

[6] Cooley R L, Koni kow L F, Naff R L. Nonlinear regression groundwater flow modeling of a deep regional aquifer system [J]. Water Resources Research, 1986, 22 (13): 1759-1778.

[J]. Journal of Applied
勘察科学技术,
997, (4):